Creating Low Carbon Cities

Shobhakar Dhakal • Matthias Ruth
Editors

Creating Low Carbon Cities

Springer

Editors
Shobhakar Dhakal
School of Environment, Resources
and Development, and Department
of Energy, Environment and Climate
Change
Asian Institute of Technology
Pathumthani, Thailand

Matthias Ruth
School of Public Policy and Urban Affairs,
and Department of Civil and Environmental
Engineering
Northeastern University
Boston, Massachusetts, USA

ISBN 978-3-319-49729-7 ISBN 978-3-319-49730-3 (eBook)
DOI 10.1007/978-3-319-49730-3

Library of Congress Control Number: 2017930816

Printed on acid-free paper

This Springer imprint is published by Springer Nature
The registered company is Springer International Publishing AG
The registered company address is: Gewerbestrasse 11, 6330 Cham, Switzerland

Forewords

Nine years ago half of the people in the world lived in urban areas, and by 2050 the world is likely to be two-thirds urban. Most people live in mid-sized cities while the large urban conurbations are growing and emerging. Cities are visible from space due to their night-time lights, and they resemble the light clusters of stars and galaxies. Like stars on matter, cities exert strong attraction on people in the hinterland where the population decreases as city populations grow.

This is one of the strong mega-trends that started with the emergence of agriculture and early civilizations that accelerated during the industrial revolution, leading to more than half of the people in the world living in urban areas. Urbanization and industrialization brought affluence to many, but a third of the global population was left behind. A quarter of the urban population lives in informal cities.

Enormous progress, despite great inequities, was only possible because the climate was stable since the onset of the Holocene almost 12,000 years ago. Global mean temperature varied less than a degree. This made agriculture possible, and thousands of years later gave rise to the industrial age.

The dark side of this process is that many people were excluded, and fossil energy became the prime mover of the world. Combustion of fossil energy leads to emissions of carbon dioxide and other greenhouse gases. Their increasing concentration in the atmosphere led to climate change, threatening the very base of human development. Recently, we have witnessed worrisome records. Concentrations of carbon dioxide have reached 400 ppmv, compared to 120 ppmv during the pre-industrial age, because humanity has emitted some two thousand billion tons of carbon dioxide during the past two centuries. We still emit some forty billion tons of carbon dioxide equivalent emissions every year. The threat of climate change is endangering human development and has resulted in worldwide calls for a paradigm change and immediate global decarbonization.

In 2015, two ground-breaking agreements were made; the unanimous adoption of the 17 Sustainable Development Goals in New York City and the ambitious Climate Agreement in Paris. The Paris Agreement calls for stabilizing temperature change to two degrees Celsius compared to pre-industrial levels, and if possible significantly less. These ambitious goals imply complete global decarbonization between 2050 and 2070.

Addressing climate change requires deep decarbonization throughout the world. Cities are a critical part of global decarbonization because of their already large direct and indirect global greenhouse gas contributions, as well as the opportunities they provide because of ongoing rapid urbanization. Cities are centers for innovation, new ideas, new policy experimentation, and large-scale infrastructural investment, adding to the climate change mitigation opportunities.

This book is very timely in addressing one of the greatest human challenges by assessing from different perspectives how and what cities can contribute toward global decarbonization. It provides fact-based knowledge and important lessons for developing low-carbon and eventually zero-carbon cities. Urban climate change mitigation knowledge and actions are fragmented along many disciplines and sectors and a book like this, which pulls from many dimensions, is a key knowledge source for actions.

Understanding the dynamics of urban development is one of the greatest scientific challenges that transcends most disciplines. Understanding deep decarbonization toward zero-carbon in urban areas is even more challenging. This book shows that the key to resolving climate change dangers is in the cities of the world.

This makes this book important and essential reading for all of those interested in the future of urbanization and deep decarbonization. I hope that it will contribute to the emergence of low-carbon cities and a sustainable future for all.

Nebjosa Nakicenovich
Deputy Director General and Deputy Chief Executive Officer,
International Institute for Applied Systems Analysis (IIASA)

There is no doubt that cities are the places where the battle against climate change can and must be won.

Cities generate as much as 70 % of all energy-related greenhouse gas emissions. They are also the places where 80 % of global gross domestic product comes from.

The capital to invest in low-carbon and clean energy technologies, the expertise of the engineers and researchers inventing and constantly improving those technologies, and the buildings and factories and transport systems that require those technologies to reduce emissions and energy consumption can all be found in cities.

A few years ago, low-carbon technologies were regarded as a niche affair. The debate was still very much focused on whether renewable energy and energy-saving technologies would ever be competitive with coal and oil. Now the debate has shifted to how fast can we make this low-carbon transition happen.

In 2015 installed power from renewable energy sources grew by over 8 %. It more than doubled in 10 years, creating over eight million jobs worldwide in the process.

Technology is not the issue anymore. As COP21 and the Paris Agreement demonstrated, it is now a matter of how can we implement low-carbon solutions to ensure that we steer clear of the worst case scenarios of climate change. Cities have a crucial role to play in this.

The challenges to such implementation are certainly formidable. To achieve a fundamental shift of global energy patterns toward a low-carbon scenario, we need many things: vast funds, in particular to help developing countries strike a balance between economic growth and sustainability; bold national policies that provide the framework for this transition to happen; capacity-building of civil servants worldwide; and in cities, particular, in finding and applying the best possible low-carbon solutions.

All this has started already. Over 600 local governments have submitted their 2020 emissions reductions plans to the carbon Climate Registry, for a total of 1 GtCO2e in emissions reductions. This figure represents a sizable percentage of the total 14 Gt gap between current Nationally Determined Contributions (NDCs) and what would be needed to be sure to keep global warming under 2 °C.

Yet, these are only a tiny fraction of the total number of cities and other local governments across the globe. Whichever technological mixes and policy instruments are chosen, monitoring and reporting progress is and will increasingly be the cornerstone of the low-carbon transition in cities.

We need more progress monitoring, more analysis of what works and what does not, and more academic inquiry to guide cities in their efforts to move away from centuries of fossil fuel-based and energy-intensive growth.

As Secretary General of ICLEI, the leading global network of 1500 local governments committed to building a sustainable future, I am therefore proud to introduce this publication, which takes stock of existing projects and trends with a multidisciplinary approach, looking at both the big picture and the small incremental steps that can be taken to proceed on this path toward sustainability.

Gino Van Begin
Secretary General, ICLEI – Local Governments
for Sustainability

Contents

Challenges and Opportunities for Transition to Low Carbon Cities

Shobhakar Dhakal and Matthias Ruth

Abstract Cities are already under pressures from several sustainability challenges. Raising income, improving livelihood, health, education, and safety, providing basic infrastructure provision such as water, mobility, energy, and housing services, and ensuring clean air, land, and water are some of them. All of these, and many others, compete for scarce financial, human, and intellectual resources. But many of these also have direct bearing on a city's carbon emissions, and investments in these services may be undermined if the contributions to and ramifications of climate change remain unchecked.

Keywords Sustainable cities • Low carbon cities • Climate mitigation • Climate adaptation • Technology change • Behavior change • Institutional innovation

Cities are already under pressures from several sustainability challenges. Raising income, improving livelihood, health, education, and safety, providing basic infrastructure provision such as water, mobility, energy, and housing services, and ensuring clean air, land, and water are some of them. All of these, and many others, compete for scarce financial, human, and intellectual resources. But many of these also have direct bearing on a city's carbon emissions, and investments in these services may be undermined if the contributions to and ramifications of climate change remain unchecked.

In recent years, cities have progressively received more attention from researchers, planners, and policy makers for the contributions they make to global climate change. Climate change is already happening and poses a serious global threat if not curtailed. The world is rapidly urbanizing, and the vulnerability of

S. Dhakal (✉)
School of Environment, Resources and Development and Department of Energy, Environment and Climate Change, Asian Institute of Technology, Pahaolyothin Road, Klong Luang, Pathumthani 12120, Thailand
e-mail: shobhakar@ait.ac.th; shobhakar.dhakal@gmail.com

M. Ruth
School of Public Policy and Urban Affairs and Department of Civil and Environmental Engineering, Northeastern University, 360 Huntington Avenue, Boston, MA 02115, USA
e-mail: m.ruth@northeastern.edu

S. Dhakal, M. Ruth (eds.), *Creating Low Carbon Cities*,
DOI 10.1007/978-3-319-49730-3_1

existing and new cities is high, given that large shares of their populations are exposed to the changes in temperatures and precipitation patterns, sea level rise, and extreme weather conditions, amongst others. Sound global climate risk mitigation requires developing low carbon cities. Cities' contribution to global CO_2 emissions are high, with direct emissions and those from electricity generation estimated to be in the range of 71–76 %.

Apparently, the world cannot address deep climate change mitigation without tapping immense mitigation opportunities in cities. As centers of economic and social activity, recipients of major infrastructure investments, and incubators of technological and lifestyle changes, cities play a major role in shaping the future of humanity. This role is only increasing as urbanization continues and as ever more people and economic assets are concentrated, often in environmentally fragile locations along coasts and hillsides, and pushed onto agricultural and wetlands and into forests.

Ecological modernization as one dominant approach, pursued globally, assures to provide services and goods to growing urban populations through growth in (urban) economies and thus expanding the technological, infrastructure, and financial resource base. Ironically enough, however, that strategy to date has largely exacerbated the root causes of the urban predicament—emissions increase with growth, fragile ecosystems are further degraded, and human-made infrastructures are overwhelmed. And in many of those places, where indeed a decoupling of economic growth from environmental impact is seen, much of it is made possible by outsourcing the problems to other locations. Islands of green behaviors in individual parts of a city or country are accompanied by higher environmental impacts in those locations from which the more affluent urbanites draw the energy and resources to maintain their lifestyles. For example, from several studies at the household level and at the city scale, it is apparent that per capita CO_2 embedded in consumption activities and the cross-boundary implications of urban consumption are enormous and surpasses direct emissions.

Rather than simply ask, for example: How can a city reduce its carbon emissions? We must also ask: What are the unintended consequences of climate action that spill over to other geographies and populations? How do actions today foster, or limit, decarbonization efforts in the future? How do these actions contribute to, or distract from, other key sustainability goals? With this volume on Low Carbon Cities we embrace many of the challenges and opportunities experienced in cities and showcase examples of policies, planning, and governance that point towards sustainable solutions. Many of these examples are, in essence, localized experiments, tuned to the broader goal of decarbonizing urban growth and development and adjusted to local conditions. The chapter on "Low Carbon Urban Design: Potentials and Opportunities," for instance, demonstrates how low carbon urban design can shape the carbon footprint of cities. Similarly, the chapter entitled "Toward Low Carbon Cities: The Chinese Experience" highlights examples of how low carbon city development in China has progressed in recent years, and provides insights on challenges and opportunities for future developments. The chapter dedicated to "Emerging Low-Carbon Urban Mega-Projects" showcases emerging low carbon mega-project such as the Masdar Eco-City and the

Sino-Singapore Tianjin Eco-city. These examples are framed by more general, conceptual perspectives that are intended to draw lessons from one setting to another. For example, the chapter on "Co-benefits and Co-costs of Climate Action Plans for Low-Carbon Cities" places low carbon strategies in the context of other urban development goals by identifying the co-benefits that may be generated by these strategies. Attention to co-benefits, so the argument, may elevate some low carbon strategies in the planning and policy process, allocating more resources to them than if co-benefits had not been considered. One such area, the water-energy-carbon nexus in urban water systems, where such synergy exists and can be harnessed is described in the chapter entitled "Optimizing Water-Energy-Carbon Nexus in Cities for Low Carbon Development", which opens up opportunities to address energy security, water security, and climate change mitigation simultaneously.

Recent studies, notably the Human Settlement Chapter of the Fifth Assessment Report (Mitigation) of Inter-Governmental Panel on Climate Change noted that, despite ambitious goals and many city level actions, the evidence of meaningful mitigation from cities yet remain elusive. Either city level climate change mitigation actions are inadequate or are at an early stage of implementation or there are many implementation problems. Stakeholders, actors, and governance are key to the success of climate action plans and social factors are central issues not to be forgotten (see the chapter on "Social Factors Affecting Low Carbon Cities").

The identification and implementation of decarbonization strategies are shaped by a diverse set of actors. These include the planners and policy makers who shape the physical and institutional boundary constraints, the investors who provide capital for the development and deployment of strategies and who, together with insurance and reinsurance companies, identify and share risks. Others include the community, citizens, and citizen groups who articulate local preferences to which, ideally, planners, policy makers, and investors respond (see the chapter entitled "Grassroots Environmentalism and Low-Carbon Cities"). However, it is not only the work of each of these groups that determines the extent to which the dynamics of cities unfold and carbon mitigation is streamlined in that realm, but also the institutional and social networks that connect them with each other at the urban scale, as well as across cities within a nation and across the globe. Under a new climate regime, new knowledge as well as actions are necessary. Universities and other research institutions, as well as government agencies, private business, and nonprofit organizations play a particular role in these networks, often serving as catalysts for change by providing knowledge products and case examples for applications of low carbon technologies and strategies.

Managing the carbon profile of cities often means stimulating and managing urban networks that consist of a wide variety of stakeholders in the public, private, and nonprofit arena. Since the global climate problem seems overly daunting to many of these stakeholders and since no individual city can noticeably influence the size of the problem or the speed at which it unfolds, the real challenge lies in the need to "think big" but to start small and to scale up fast. New technologies, such as access to and utilization of the large data streams describing activities of individuals, business, and agencies in the urban system can offer actionable information and new opportunities at unprecedented granularity (chapter "Big Data, People and Low Carbon Cities"). Innovations in and deployment of green building and clean energy technologies can lay the footprint for lower carbon emissions (chapters on "Emerging Low-Carbon Urban Mega-Projects," "Low-Carbon Urban Infrastructure" and "Low-Carbon Waste Management"). But these technologies alone will show limited impact if, for example, they are not scaled up and not adequately supported by financial instruments, many of which are as innovative as the technologies themselves (see e.g., the chapter on "Low Carbon Cities: The Chinese Experience"), and if they are not combined with the social factors and political will for implementation (see chapters entitled "Grassroots Environmentalism and Low-Carbon Cities," "Social Factors Affecting Low Carbon Cities" and "Eco-Districts as a Transition Pathway to Low-Carbon Cities"). In short, deployment of low carbon strategies requires more than development of low carbon technologies. Instead, it requires also innovation in the legal, institutional, and financial world, and new governance structures that allow low carbon technologies and behaviors to unfold and flourish.

Transition to low carbon cities must enable cities to begin to operate in different ways and use new performance measures by which their progress and development are assessed (see the chapters on "Energy Consumption and Emissions Assessment in Cities: An Overview," "Managing Greenhouse Gases Emissions in Cities: The Role of Inventories and Mitigation Actions Planning," "Potential Transformation Pathways Towards Low Carbon Cities: The Big Picture"). Instead of traditional economic growth goals, stakeholders must drive a city towards a decoupled growth-environment scenario. A change in the mind-sets of social actors, planners, and policy makers will be required by which globally agreed-upon targets on emissions reductions are met by diverse, locally fine-tuned experiments. All this, in turn, calls for a change in focus from plans and policies to actual implementation, from incremental changes to the transformative ones, and towards a long-term enabling environment. It is our goal to contribute with this volume to that shift in mind-sets, strategies, and actions.

Big Data, People, and Low-Carbon Cities

Paul Fleming

Abstract Information- and communications-technology systems are now collecting and analysing previously unimaginable amounts of data in cities. Big data and the "internet of things" are allowing communications between city infrastructure (buildings, transport systems, equipment, and appliances) and people. For example, data are collected by a wide range of public and private organizations, and people are sharing their own local weather and local solar-electricity production. This chapter discusses how cities can use the information from this increasing volume of data to decrease their carbon emissions, help create new employment opportunities, facilitate future infrastructure investment, and more effectively engage with their citizens.

Keywords Energy management • Low-carbon city • Future city • Smart grids • Community engagement • Citizen data • Data analysis

Key Terms

Big Data. The large volume of structured and unstructured data that is now available from buildings, transport systems, people, etc. It refers to data that are too big to be analysed using traditional data-processing techniques.

The Internet of Things. The representation of uniquely identifiable objects on the Internet. A medium for transferring data between a data source and the processes that monitor those data sources. Enabling things to communicate with each other.

Virtual power plant. A collection of distributed energy-supply technologies (fossil fuel or renewables) that can be virtually managed as "one single power plant".

Smart grid. An energy network that is able to optimise energy supply with demand by monitoring and controlling energy supply, storage, and demand. It decreases peaks in demand by remotely switching off equipment or shifting the use of that equipment to another time period.

P. Fleming (✉)
Institute of Energy and Sustainable Development, De Montfort University, Leicester, UK
e-mail: pfleming@dmu.ac.uk

S. Dhakal, M. Ruth (eds.), *Creating Low Carbon Cities*,
DOI 10.1007/978-3-319-49730-3_2

Smart city. A city that uses information and communication technology to help improve governance, mobility, carbon emissions, and peoples' quality of life. It uses big data and Internet of Things to manage city infrastructure and to provide feedback from, and dialogue with, city stakeholders and citizens.

Digital divide. Inequality between people and cities in terms of their access to, and knowledge of, information and communication technology.

1 Introduction

Cities have been taking action to decrease their greenhouse-gas emissions for many years. Organisations—such as Energy Cities [1], Fedarene [2], ICLEI [3], Climate Alliance [4], the Covenant of Mayors [5], the C40 Cities Climate Leadership Group [6], Eurocities [7], and the OECD [8]—all possess a wealth of information on how to decrease greenhouse-gas emissions at the city scale. Traditionally, local and regional emissions data were derived from annual records of electricity, gas, oil, and solid fuel use, which account for >80 % of emissions [9]. City energy managers then focused on specific programmes to decrease their emissions and estimate the potential reductions [10]. Sub-hourly data from utility meters and building energy-management systems now provide energy-related data almost in real time. Traffic-management systems provide real-time traffic information and make it available on the Internet. The Internet of Things has resulted in many devices being Internet enabled, thus allowing for two-way communications with "things" that are now being controlled rather than just measured. This brings with it the potential for "smart" systems. However, these smart systems must be robust to cope with potential disruptions. For example, they may need to include local data storage to address potential cascading failures when electricity network interruptions disrupt the performance of sensors and communications networks. Such disruptions could undermine the reliable operation of future smart networks as well as that of future smart cities. An example is the city of Lancaster, UK, where flooding caused the electricity system to be unavailable for 4 days resulting in a loss of data communications and problems for people living and working in the city [11].

Future cities will not operate the way cities did in the past, with centralised power supplies. Separate electricity, gas, heat, and transport networks will not deliver ambitious energy- and carbon-reduction measures over the long term. These networks will need to be integrated, expanded, and made part of an effective ICT (Information and Communications Technology) network to collect, share, and analyse data to provide information required to optimise their operation. The networks will also need to have an element of both electrical and thermal storage to cope with "peaks" and "troughs". Information technology can then provide the communication between city energy supply, energy demand, and energy storage, thus enabling intelligent demand-management measures to optimise efficient energy supply, storage, and use [12]. Transport will also be part of this integration, and electric vehicles will have the potential to act as electrical sources for energy storage for powering buildings as well as vehicles. There will be not only individual heat and power generation but also individual energy storage in homes and businesses. Homes and businesses will no longer be just consumers, they will be energy producers as well. The city energy supply will therefore be much different than it has been in the past. Traditional, highly centralised power generation will be replaced with new decentralised systems that will operate at a very local level. These systems will match the intermittent supply to the demand by using demand shifting and storage. Data analysis will be a key part of the successful operation of such decentralised systems. People and businesses should then have greater engagement with heat and power suppliers because they are producers (and generating an income) as well as consumers (consuming at different prices at different times of the day with new tariff structures). Complex control software must be available to optimise such different supply, demand, and storage options at scale. Many local producers are already sharing their data on solar-electricity production on Web sites (e.g., www.pvoutput.org) to share, compare, and monitor live photovoltaic-electricity generation. These new decentralised systems will help overall system resilience; however, many legal, ethical, and operational challenges must be overcome before they will be widely adopted.

In the past, city managers had relatively limited engagement with users of the city's day-to-day services. They had to arrange specific public meetings or surveys to get feedback on how effectively systems were operating as well as any proposals for new city infrastructure. However, we now have cities with dedicated telephone and Web-based "hot lines" and people using social media, again allowing for two-way communications, to tell the world their views on a wide range of things—including low-carbon cities. Cities can now use the data from city infrastructure (buildings, transport, etc.) and data from people (by way of Web sites and social media) to help deliver a low-carbon city. They can more rapidly identify when things are going wrong—such as power outages or internet failures as well as traffic delays—so they can maintain smooth operation of the city.

An integrated, multi-disciplinary approach, in partnership with other stakeholders, is needed to achieve the deep cuts in carbon emissions needed to meet international targets. For example, achieving a 5 % reduction in energy consumption in a building can be delivered through addressing one single technology. Either providing additional thermal insulation, improving heating or cooling systems, or improving building-control systems should result in >5 % savings. However, our challenge is for western cities to achieve savings on the order of 80 % to keep the global temperature increases to <2 °C as agreed at the Paris COP in 2015 [13]. To do this, it is not sufficient to determine a solution from one single technology or discipline. An interdisciplinary approach must be employed using several technologies (thermal insulation, heating and cooling systems, lighting and ventilation systems, and improved controls), and then we must go beyond the technology itself to address how people effectively make use of these technologies in their daily lives. A comprehensive, whole-systems approach to energy and carbon emissions is therefore needed at the city scale.

Information- and communication-technology systems now allow for the collection and analysis of previously unimaginable amounts of both city-infrastructure and people-related data. Data are therefore available to help estimate people's carbon emissions almost in real time. The challenge is how we move from this "big data" to "big knowledge" and how we develop better approaches to data visualisation. This will help cities to share data and compare the performance of different lifestyles, homes, businesses working practices, etc. These data could also help businesses develop new products and services based on more detailed and relevant information about carbon emissions. Services reflecting this concept have been developed, for example, by Uber and BlaBlaCar for mobility services. Overall these data can then be used to manage the "carbon system" of the city. They can be used to enable a city to operate a smart-electricity grid, smart-heat grid, and smart-transport grid as well as help create new employment opportunities and facilitate inward investment and better engagement with, and participation from, the public.

2 What Data Exist

Data are routinely collected regarding energy supply (electricity, heat, and transport) and fuel type (fossil-fuel heat, fossil-fuel electricity, renewable heat, renewable electricity, and transport-related fuels). We can measure this supply by different types in real time. Smart meters are available in people's homes. Data are also collected, by way of automatic meter readings of the energy demand in buildings and industry (electricity and heat in buildings) and transport (electricity and oil). Building energy-management systems (BEMS) collect data on temperature, lighting, ventilation, etc., in different rooms in different buildings. Air quality–monitoring stations collect local weather and air pollution data. Process-control systems record data on industry, and transport-management systems collect data on vehicle movements, passenger loading, congestion, car-park occupancy, electric-vehicle charging points, bus and train times, etc. Car manufacturers collect data on engine performance, speed, location, etc., which can be used to provide feedback to drivers so they change driving styles to decrease emissions. Remote sensing by satellite enables the monitoring of local air pollution in real time, and examples exist of waste being tracked from the recycling point to eventual reuse and disposal.

Geographical Information Systems allow this information to be displayed as maps, and new techniques are being developed to help people visualize these data. People can share information about themselves and their "happiness". All of these data can be collected and analysed. Systems are now moving beyond monitoring and displaying information to also controlling things. Washing machines, cookers, heating systems, and televisions can now be switched on and off remotely by way of mobile phones. Machines can communicate with machines without human intervention. However, gaining access to these data at meaningful geographic and temporal resolutions and ensuring consistent quality can still be a problem.

All of this now makes it possible to better predict energy use in cites and to offer people incentives to both decrease and time-shift their heat, light, and power needs. Cities are installing improved information and communications technologies, and open-source approaches are allowing access to large data sets. They are now using these data to become intelligent future cities and citizens. Communities and individuals are self-reporting and sharing their data. These data can then be analysed to act as a proxy for population density or to highlight future development options for communities. However these data bring ethical, security, and privacy issues. People are become more concerned about how their data are used. Whose owns the data? What is proprietary data, and what is publicly available? Legal, ethical, privacy, and security issues still remain to be overcome.

3 What Data Do Not Exist

Using data that currently exist, it is possible to estimate the annual carbon emissions at a city level, but this cannot yet in be done in real time or in terms of understanding the detail of citizens' behaviours. The challenge is taking these existing data, obtaining further data, and making use of the information provided by these data. We do not yet have the quality and quantity of data to provide a true city-wide picture in real time. Data are not available uniformly; rather, there are concentrations of data. For example, there are substantial energy-related data for buildings as well as transport-related data through traffic-management systems, car parks, and bus and train time information. However, there are still minimal data available on carbon emissions associated, for example, with food procurement and waste. How to deal with the variations in availability and quality of data is a key issue for cities not just to address the digital divide but also for the research community to develop new algorithms to help close the gaps in data and knowledge.

A further key issue is the collection, storage, and processing of these data. Substantial interoperability issues associated with data acquisition still remain. Data are collected in different formats over different time periods at different locations, and they are of varying quantity and quality. For example, whilst common standards for energy-related data exchange have been suggested [14], they have not yet been widely adopted. More data are becoming available all the time with the Internet of Things and open-source approaches. However, the issue is not just the existence of data but the relevance, quality, and usefulness of these data as well as how to address the associated security, privacy, and ethical issues. Maintaining quality of data over time still remains a major issue.

4 What More Is Needed?

First, as we collect increasingly more data, we need better institutions to oversee their collection and management. This is needed not only to ensure quality but also to reassure the public that ethical, privacy, and security issues have been addressed. Second, we need better ways of extracting useful information from these data, information to help city **decision makers** improve the management and future infrastructure of our cities. Finally, additional real-time data are needed to help provide more information about carbon emissions from infrastructure as well as from people. These data will then need to be collected, processed, stored, analysed, and presented to users. Interoperability, ethics, privacy, and security issues must also be resolved. We need infrastructure-related data on buildings, transport, food, and waste as well as people-related data in terms of their attitudes and behaviours. Using these infrastructure and people data, "typical" citizens in different house types—with different jobs and taking part in different leisure and social activities—could be constructed. These virtual citizens could then be compared with people's similar lifestyles and advice and guidance given on how to decrease emissions through making changes in lifestyles. This reduction could then be measured and quantified.

As more data become available, we will need to develop new techniques to analyse them, E.g., new approaches to presenting these data in the most appropriate way to different users including city leaders, city mangers, city businesses, and citizens. For example, city leaders may not have traditionally used this level of data before. They will need to have information derived from the data in different forms than that needed by city managers. They will also be very acutely aware of ethical, privacy, and security issues that can cause significant reputational damage if not addressed effectively from the outset. In addition, we still must overcome the digital divide and address regular public concern over the release of heath and other data to private companies. Ethical, security, and privacy issues will therefore be key issues to address in parallel with other data-quality, -management, and -visualisation issues.

5 How to Use Data

Data visualisation is key to making better sense of data. We must know how to present the most appropriate outcomes from the complex analysis of large data sets to city practitioners and the public so that they are able to use it effectively. This ranges from providing simple smiley faces to represent electricity gas- and water-consumption trends in buildings [15] to city dashboards [16] to complex visualisation of multi-layer urban-data platforms for citizen and policy-maker exchange as in Live Singapore [17] (http://senseable.mit.edu/livesingapore/index.html). The growing areas of data mining and data visualisation will continue to be key future research topics. Such an approach is being adopted by the European Commission's

Collective Awareness Platforms (https://ec.europa.eu/digital-single-market/en/collective-awareness), which are supporting environmentally aware, grassroots processes and practices to share knowledge, achieve changes in lifestyle, production, and consumption patterns, and set up more participatory democratic processes all using city-wide data to engage with different users. An example is the Political and Social Awareness on Water Environmental Challenges (POWER [http://www.power-h2020.eu/overview/]) project, which engages policy-makers, professionals, and citizens in water-related issues identified by the European Innovation Partnership (EIP) Water Action Group City Blueprints, thus improving the implementation capacities of cities and regions (http://www.eip-water.eu/City_Blueprints). This project brings together public bodies, private bodies, and communities to share knowledge and data by way of a digital social platform to address specific local water-related issues through building on the analysis of local city data.

Large amounts of data—including data from both city infrastructure and citizens available in real time—are therefore being collected, stored, analysed, and presented to potential users. The infrastructure, in terms of networks and smart grids (smart heat, smart electricity, smart transport, and smart energy storage), is being expanded. In terms of information from the average citizen, people share about the following:

- where they are, how they travel, and how much electricity, gas, and water they are using in their homes or at work;
- what temperatures they maintain in their home and working environments; and
- what information they are sharing with other people on social media and other Web sites.

In addition, businesses share their energy efficiency and benchmark these data with those of similar businesses in other countries. The information gathered from these data can be used to help deliver and manage a low-carbon action programme. Information from the analysis of these data thus inform the delivery of large-scale low-carbon interventions such as a district-wide or community-wide refurbishment projects, new low-emission vehicle programmes, air-quality action zones, refurbishment of non-domestic buildings, new business opportunities and models, etc. It also allows for the ongoing monitoring of these schemes and provides evidence to convince decision-makers of the business case for further investment.

These data can inform a city-wide carbon-management system that can measure and control the operation of different city infrastructures. Machine-to-machine communication will deal with the control and operation of some systems. It can provide people with real-time feedback on carbon emissions as well as information about emissions from homes, businesses, and modes of transport. Such emissions are related to different peoples' lifestyles. It can form part of a city dashboard by describing current city-wide emissions. It will enable cities to move beyond simple monitoring to introduce control, management, public engagement, and pubic feedback. However, data quality, long-term storage and management, ethics, interoperability, and privacy issues remain to be resolved.

6 Conclusions

Big data and the Internet of Things are now providing the opportunity to help deliver a low-carbon city. The amount of data and the number of things connected to the Internet are increasing. Future cities cannot operate the way they did in the past with separate electricity, gas, heat, and transport networks. These will not deliver the internationally agreed-upon energy- and carbon-reduction targets. The networks will need to be integrated as well as have an element of both electrical and thermal storage with the electrical storage including electric vehicles where the stored electricity is used to power buildings as well as the vehicle. Information technology provides communication between energy supply, energy demand, and energy storage, thus resulting in intelligent supply- and demand-management measures to optimise energy supply, storage, and use. Homes and businesses will be producers as well as consumers of energy. From an energy perspective, we now (in theory) have access to the real-time data to develop a virtual power plant for a city. It is conceivable to link a smart-electricity network with a smart-heat network and a smart-transport network having both electrical as well as thermal demand response and storage.

Analysis of the vast amount of data and the information from this analysis is then used to inform the "management" of the city. Much of this ongoing, daily management is performed automatically, machine to machine, by way of software. Specific fault correction and efficiency improvements are performed by energy-service companies. Information from the analysis of these data will enable cities to better understand the needs of their citizens through a more regular dialogue based on citizens' real-time carbon emissions to move beyond simple monitoring to intelligent control. However, ethical, privacy, and security issues need to be overcome, and the digital divide could prevent some citizens from realising the full benefits of a smart city. The key is ongoing monitoring and evaluation at a city scale. From an energy manager's perspective, this involves moving from monitoring, target setting, and investment in buildings to monitoring, target setting and investment in cities. It means moving toward smart cities.

7 Summary

- Data are available to help estimate and manage carbon emission at the city level. The amount of data is increasing, and the relevant data can be harnessed to decrease carbon emissions in our cities.
- Both data from city's infrastructure—homes, businesses, and transport—and data about people living, working, and studying in cities are available.
- Security, privacy, and ethical issues, as well as the digital divide, must be addressed from the outset.
- Citizens, communities, and businesses are sharing their data.

- Data can help justify low-carbon investment opportunities, and ongoing city management can be delivered through virtual power plants and smart grids with energy supply, storage, and demand management.
- Using these data, it is possible to develop evidence-based policy regarding city efficiency. The main goal is to evaluate a city's performance in decreasing greenhouse-gas emissions and to evaluate citizens' responses to this policy.

References

1. Energy Cities: http://www.energy-cities.eu/
2. Fedarene: http://www.fedarene.org/
3. ICLEI: http://www.iclei.org/
4. Climate Alliance: http://www.climatealliance.org/
5. Covenant of Mayors: http://www.covenantofmayors.eu/index_en.html
6. C40 Cities Climate Leadership Group: http://www.c40cities.org/
7. Euro Cites: http://www.eurocities.eu/
8. OECD: Urban Energy Handbook—Good Local Practice. OECD, Paris (1995)
9. Shackley, S., Kersey, J., Wilby, R., Fleming, P.: Changing by Degrees: The Potential Impacts of Climate Change in the East Midlands. Ashgate, Aldershot (2001)
10. Fleming, P., Webber, P.: Local and regional greenhouse gas management. Energy Policy. **32**, 761–771 (2004)
11. Kemp, R.: Living Without Electricity. Royal Academy of Engineering, London (2016) http://www.raeng.org.uk/publications/reports/living-without-electricity
12. Boait, P., Ardestani, B.M., Snape, J.R.: Accommodating renewable generation through an aggregator-focused method for inducing demand side response from electricity consumers. IET Renew. Power Gen. **7**(6), 689–699 (2013)
13. UNFCC: http://newsroom.unfccc.int/unfccc-newsroom/finale-cop21/
14. AMON: http://amee.github.io/AMON/
15. Stuart G., Fleming P. (2014) Smart energy performance indicators for live historical and normative feedback systems. Improving Energy Efficiency in Commercial Buildings (IEECB) conference 2014, Frankfurt, pp. 400–414
16. Future City Dashboard, Glasgow: http://dashboard.glasgow.gov.uk/
17. Kloeckl, K., Senn, O., Ratti, C.: Enabling the real-time city: LIVE Singapore! J. Urban Technol. **19**(2), 89 (2012)

Paul Fleming is a professor of energy management and the Director of Sustainable Development, Institute of Energy and Sustainable Development, De Montfort University, Leicester, UK. His research focuses on energy efficiency, renewable energy, and greenhouse gas emission reductions at the local to regional levels, in particular on the analysis of energy-related data to help achieve deep cuts in emissions. This works addresses the technical, non-technical, and public-engagement aspects of delivering low-carbon cities and improving citizens" quality of life.

Co-benefits and Co-costs of Climate Action Plans for Low-Carbon Cities

Matthias Ruth, Sanchari Ghosh, Sahar Mirzaee, and Nancy S. Lee

Abstract City-level climate action plans are often designed to address specific issues such as to cut GHG emissions from traffic congestion. The benefits from such plans would include the direct effects of reducing contributions by cities to global atmospheric GHG concentrations. Additional benefits may be present in the form of energy savings, reduced air pollution, improved public health, and many more. The presence of such co-benefits (and of co-costs) may affect the rank-ordering of particular actions when they are compared against each other. This chapter provides a structured approach to the assessment of co-benefits and co-costs, and their implications for selection among climate actions. Using network analysis we assess existing urban climate action plans from around the world, focusing on the notion of co-benefits and co-costs. We find notable similarities and differences in the way co-benefits and co-costs guide urban climate plans, and we offer guidance for the social discourse on prioritization of strategies.

Keywords Mitigation • Adaptation • Co-benefits • Co-costs • Urban climate

1 Introduction

Importance of City-Level Climate Mitigation and Adaption

The Fifth Assessment Report of the IPCC (2014) [1] has pointed out that there is a high probability that around half of observed climate change is attributable to anthropogenic sources of greenhouse gas (GHG) emissions over the period 1951–2010. Cities have become dominant sources of GHG emissions from energy consumption [2] as well as the victims of climate change induced vulnerabilities due to the presence of

M. Ruth (✉)
School of Public Policy and Urban Affairs and Department of Civil and Environmental Engineering, Northeastern University, 360 Huntington Avenue, Boston, MA 02115, USA
e-mail: m.ruth@northeastern.edu

S. Ghosh • S. Mirzaee • N.S. Lee
School of Public Policy and Urban Affairs and Resilient Cities Lab, Northeastern University, Boston, MA 02115, USA

S. Dhakal, M. Ruth (eds.), *Creating Low Carbon Cities*,
DOI 10.1007/978-3-319-49730-3_3

factors such as expanding urban heat islands, growing urban populations near coastlines, exacerbated levels of pollution, and rising population density [3, 4]. With growing rates of urbanization [5] and often lacking national action, increased attention is placed on city-level climate change mitigation and adaptation plans, most notably in the larger cities such as New York and Chicago in the USA, London, England, Cape Town, South Africa, Shanghai, China, and Melbourne, Australia, but also smaller cities around the world like Livermore and Alexandria in the USA or Freiburg in Germany. Cities are developing their own, specific action plans targeted at cutting their greenhouse gas emissions, reducing the impacts of climate-induced risks, and implementing measures to enhance community-level resilience to counter these risks. Their policies range from promoting efficiency in energy use in the transportation, agriculture, and industrial sectors, to building green infrastructure and automated wastewater recycling facilities, building flood control infrastructure, and establishing solid waste treatment and recycling facilities.

Significance of Including Co-benefits and co-costs in City-Level Climate Action

City-level climate action plans are often designed to address specific issues such as to cut GHG emissions from traffic congestion. The benefits from such plans would include the direct effects of reducing contributions by cities to global atmospheric GHG concentrations, *plus* the co-benefits that are generated in terms of energy savings, reduced air pollution, and improved public health [6–9]. While the direct benefits for global climate are diffuse and will occur over the long term, co-benefits accrue more immediately and directly to the city [10] and thus may potentially provide additional impetus for climate action at the city level.

> Failure to include all co-benefits will lead to underestimation of the full benefits of climate change mitigation. The majority of co-benefits identified in climate action plans are health-related. Health co-benefits refer to the reduction in health-related costs due to improvement in air or water quality as a result of climate change mitigation policies.

2 Economic Argument for Including Co-benefits and Co-costs

Optimal Abatement Levels Including Direct Benefits and Co-benefits

To define and capture co-benefits and co-costs requires systematic treatment of system boundaries in space and time, and commensurability of these costs and

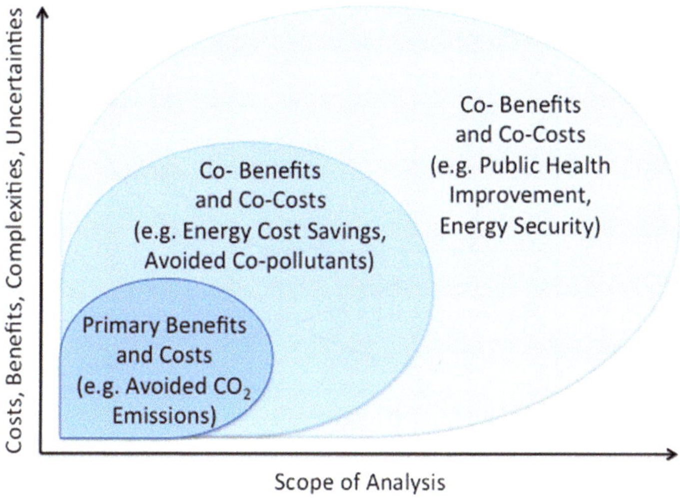

Fig. 1 Primary benefits and co-benefits

benefits with other performance measures. For example, monetary values may be imputed for public health costs of declining indoor air quality that results from reduced ventilation of living spaces. Changes in comfort, for example, are more difficult to quantify in monetary terms, as are the increases in national energy security that result from reductions in domestic heating and cooling requirements. As the scope of analysis increases, the ancillary effects of policies and investments become more diffuse, and the ranges of estimates for costs and benefits broaden (Fig. 1). This calls for the establishment of standards and agreed-upon procedures for such accounting, at least for two main reasons. First, to the extent that total benefit and cost measures are used to guide policy, selective inclusion or exclusion of (co-) benefits and (co-) costs will bias decision-making [11]. Second, total ancillary benefits may exceed primary benefits [12], and even where they do not, their inclusion may help magnify the primary benefits [13] and can alter the rank-ordering of preferred policy choices [6].

Figure 2 shows the standard case of damage costs, $D\ (Z)$, declining for increasing levels of emissions reduction, Z. The costs of carrying out those reductions (the abatement costs), $R(Z)$ are usually assumed to rise both monotonically because avoiding the next unit of pollution will be ever harder to do. From an economic standpoint, the optimal abatement level Z_0 is obtained when the benefits from reducing emissions by the next unit just cover the costs of avoiding that next unit from being released into the environment. An excellent illustration of such abatement cost curves is the global abatement cost curve [14], which shows the abatement cost of 1 ton of carbon dioxide across various sectors through 2030, relative to a business-as-usual (BAU) development.

The presence of co-benefits shifts the emissions cost curve $R(Z)$ downward, for at least some level of emissions reductions, thus resulting in optimal abatement

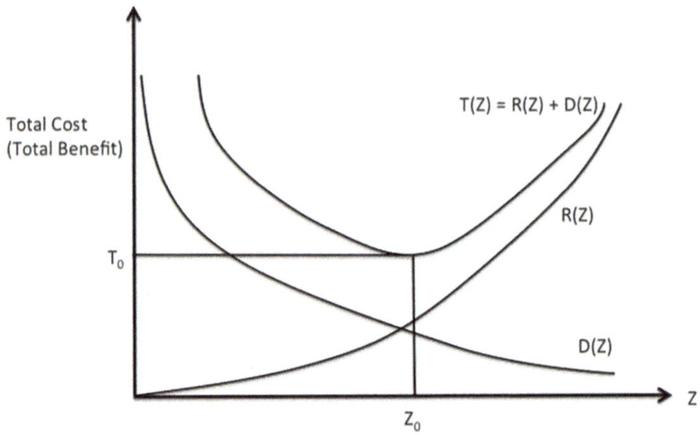

Fig. 2 Standard damage cost and abatement cost profiles

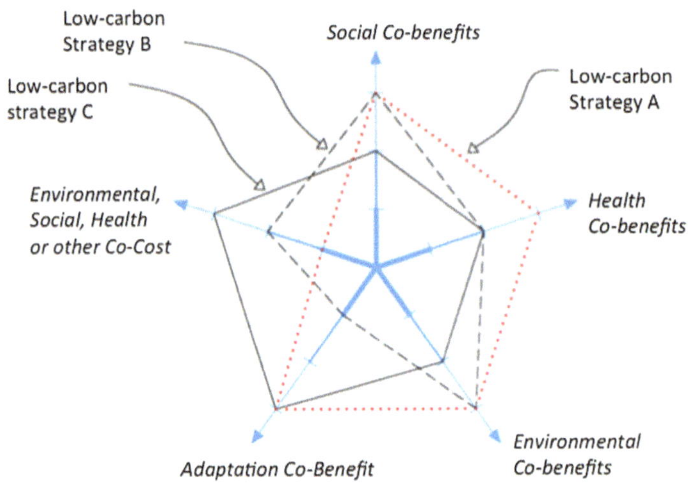

Fig. 3 Schematic representation of multiple decision-making criteria for alternative low-carbon strategies

levels larger than Z_0, while co-costs shift $R(Z)$ up, thus making emissions reductions more costly and leading to abatement levels below Z_0. The increased complexity and uncertainty associated with measuring co-benefits and co-costs, however, will make it very difficult to identify with standard economic instruments, at any meaningful level of accuracy, "optimal" low carbon strategies.

Instead, social discourse will need to help decision-makers prioritize among the many mitigation options and their respective co-benefits and co-costs. To do so, various methods exist to structure the disparate information and associated uncertainties about co-benefits and co-costs, such as the visualizations of multiple decision criteria schematically shown in Fig. 3 for five stylized categories of co-benefits and co-costs (see, for example, Dias and Domingues [15], for a critical discussion and examples). Of course, many of these co-benefits and co-costs have been already alluded to in the climate action plans of cities, albeit often in less structured ways. The following sections of this chapter review the extent to which these plans do consider ancillary benefits and costs, and how those are perceived to be connected to low-carbon goals.

3 Representative Cities and Their Climate Mitigation Plans

The map in Fig. 4 depicts the cities selected for this study spanning all continents except Antarctica. It is evident that the cities in the developing nations are having explosive rates of population growth as compared to the cities in the developed countries.

Cities in Developed Nations

The selected cities in the developed nations comprise 9 out of the 20 most densely populated cities in the USA (Chicago, Boston, Columbus, Los Angeles, Phoenix, San Diego, New York, San Jose, and Austin) and eight select cities in Europe, which have an ongoing city climate action plan (London, Rome, Paris, Malmo, Stockholm, Rotterdam, Madrid, and Hamburg). Our sample also contains the four

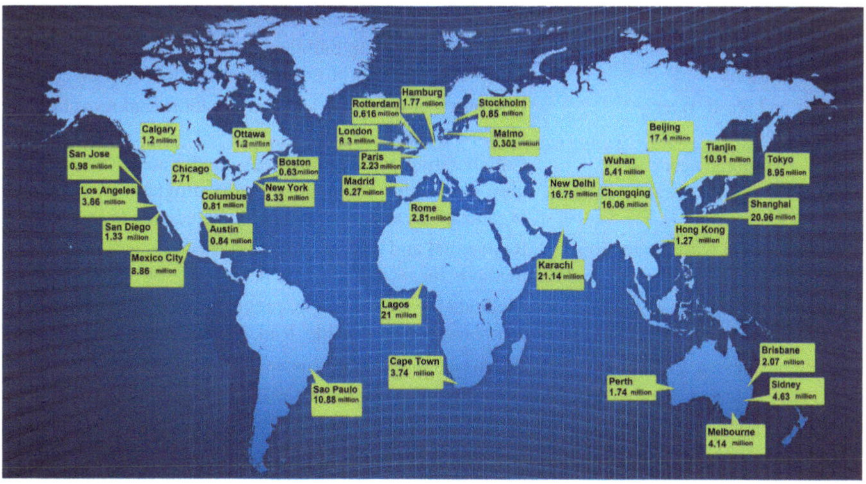

Fig. 4 Cities with climate action plans and their population

biggest cities in Australia (Sydney, Melbourne, Brisbane, and Perth), Tokyo in Japan, and Calgary and Ottawa in Canada.

Cities in Developing Nations

Eleven cities in the developing nations constitute a comprehensive list of cities with ongoing or recently initiated climate action plans among the emerging and less developed economies. These cities are Hong Kong, Beijing, Guangzhou, Shanghai, Tianjin, Wuhan, and Chongqing in China, with the last four constituting the mega-cities with populations exceeding ten million [15], Cape Town and Lagos in Africa, Mexico City and Sao Paulo in South America, and New Delhi and Karachi in South Asia.

Climate Action Categories

The review of city-level climate adaptation plans reveals 14 broad categories of actions, summarized in Table 1. While each city action plan in the developed or developing countries is predominantly designed in the context of climate change issues of that particular city, notable structural differences exist. For example, developed-country cities like New York, Chicago, Boston, and London have been

Table 1 Action categories

Action category	Description of category
Transportation technology	Improvement in fuel efficiency and transportation infrastructure; emphasis upon the use of public transport
Green buildings, systems, and operations	Energy efficiency in building construction and operations and encouragement of green infrastructure
Water and wastewater management	Efficiency in water use and conservation, and recycling of wastewater
Public relations: agencies and community engagement	Collaborating with communities on climate action strategies, developing community energy programs and green organizations and promoting climate education in schools
Recycling and waste reduction	Residential and commercial solid waste reduction and methods for recycling waste matter
Outdoor and indoor air quality	Methods for minimizing exposure to air pollutants such as ozone, carbon dioxide, lead, and particulate matter
Green businesses	Promotion of green businesses like eco-industry, eco-agriculture, and green service industry which can facilitate employment generation
Energy conservation	Promotion of energy efficiency standards for new buildings and commercial facilities, appliances and the water sector

(continued)

Table 1 (continued)

Action category	Description of category
Renewable energy	Encouraging the use of solar and wind energy and green electricity for clean air, and conservation of fossil fuel
Land use and food systems	Urban design for energy conservation, tree planting and establishment of community gardens and adoption of sustainable land use and overall promoting smart growth policies for cities
Behavior change	Raising public awareness of climate change and motivating them to conserve energy, use more of public transport, conserve water and reduce waste, and promote green infrastructure
Research, studies, and assessments	Establishing centers and encouraging studies on the impacts of climate change and how to build resilience
Severe weather forecasting/ warning	Developing and implementing warning systems and response plan for combating severe weather
Healthcare: climate change related diseases	Promoting awareness on climate change related diseases, developing emergency health measures for them and enhancing the capacity of medical personnel to effectively manage climate change related health diseases

forerunners in devising strategies that involve community participation and using low-carbon strategies as part of an economic development agenda to create "green" jobs. Cities like Sao Paolo, Shanghai, Mexico City, and Cape Town exhibit an evolving climate action framework with broad perspectives on the interactions among energy, water, building infrastructure, transportation, and land use, but community representation as part of climate action strategies remains low in many of these cities, especially those in Asia.

4 City Climate Action Plan Analysis

Differences in Action Categories and Their Frequencies

Table 2 illustrates the ranking of categories in each continent based on the frequency of the categories mentioned in their respective action plans. The action categories have qualitative differences since the goals, methods of implementation, and infrastructure vary among various cities around the world. However, the frequency ranking is carried out for two main reasons. One, it is a first step towards identifying which mitigation strategies have been perceived to be of high importance by city climate action planners and policy makers. Secondly, knowledge of this ranking will generate initial discussions on the extent to which cities in different regions prioritize climate action policies and allocate their resources towards such policies.

Table 2 Frequency ranking of action categories

Rank	North America	South America	Asia	Africa	Europe	Australia
1	Transportation technology	Transportation technology	Green buildings, systems, and operations	Land use and food systems	Green buildings, systems, and operations	Transportation technology
2	Water and wastewater management	Water and wastewater management	Transportation technology	Transportation technology	Transportation technology	Behavior change
3	Green buildings, systems, and operations	Land use and food systems	Land use and food systems	Water and wastewater management	Energy conservation	Land use and food systems
12	Green businesses	Green businesses	Outdoor and indoor air quality	Outdoor and indoor air quality	Outdoor and indoor air quality	Outdoor and indoor air quality
13	Severe weather forecasting/warning	Research studies and assessment	Severe weather forecasting/warning	Green businesses	Severe weather forecasting/warning	Green businesses
14	Healthcare: climate change related diseases	Behavior change	Healthcare: climate change related diseases	Severe weather forecasting/warning	Healthcare: climate change related diseases	Healthcare: climate change related diseases

Improved transportation technology; green buildings, systems, and operations; and water and wastewater management with land use and food systems receive high priority in Africa, Asia, and Latin America. The relevance of land use and food in Africa is not unexpected since African countries are prone to frequent droughts and food scarcity. In contrast, the emphasis on improved transportation technology; green buildings, systems and operations; and water and wastewater management imply that the consequences from improved air and water quality and health seem to be predominant driving factors for these action plans.

Identification or Recognition of Co-benefits and Co-costs

A strength of the climate action categories in the city adaptation plans lies in determining and sometimes quantifying the potential co-benefits and co-costs from implementation of each plan. Among the 36 cities with climate action plans, 24 are in developed countries and 20 cities identified co-benefits in their plans; 12 are in developing countries and seven cities identified co-benefits. However, very few cities have estimated the magnitude of co-benefits or co-costs associated with each action category (Table 3).

> A full network may have dense interconnections between certain actions and their consequences. The density represents more connectivity between and among actions and their ancillary benefits or co-benefits.

A network analysis is a convenient way to represent the detailed coupling of climate actions and co-benefits that are associated with GHG emission reductions. In the absence of actual estimates of the magnitudes of co-benefits, the network

Table 3 Summary of cities identifying co-benefits in climate action plans

Cities that mention co-benefits in their climate action plans					Cities that do not mention co-benefits in their climate action plans	
Developed			Developing		Developed	Developing
Boston	Los Angeles	Rotterdam	Beijing	Mexico City	Austin	Karachi
Brisbane	Madrid	San Diego	Cape Town	Sao Paulo	Malmo	
Calgary	Melbourne	San Jose	Chongqing	Shanghai	Paris	
Columbus	New York	Sidney	Delhi	Tianjin	Rome	
Hamburg	Ottawa	Stockholm	Guangzhou	Wuhan		
Hong Kong	Perth	Tokyo	Lagos			
London	Phoenix					

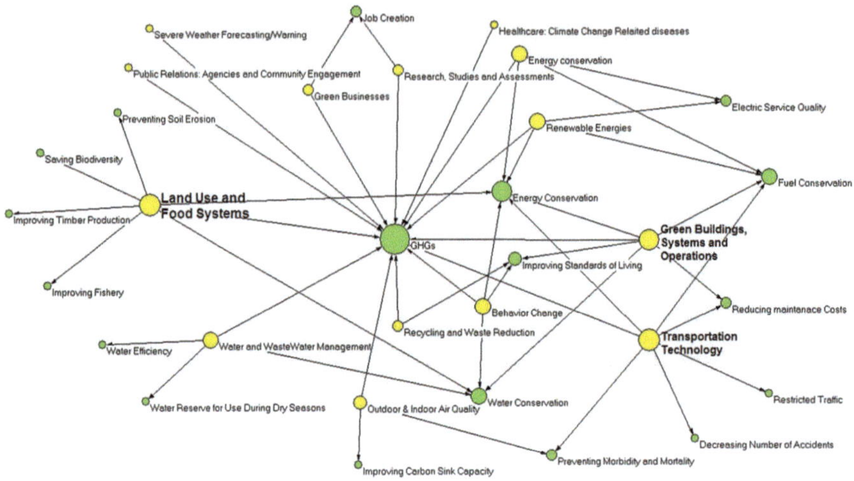

Fig. 5 Network diagram for US city co-benefits from GHG emission reductions

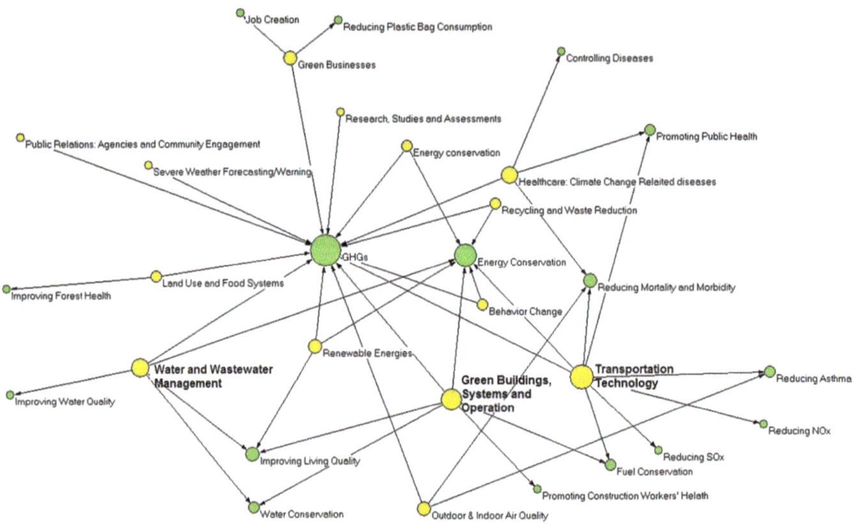

Fig. 6 Network diagram for Asian city co-benefits from GHG emission reductions

identifies the number of ancillary benefits from action plans targeted at lowering carbon emissions. Figures 5, 6, and 7 present the connections among co-benefits and GHG emissions reductions in US cities, selected Asian cities (except Tokyo, because it markedly differs from cities in developing Asian continent), and African cities. Climate action plan categories are highlighted in yellow while the co-benefits appear in light green. Noticeably underrepresented in each of these networks are

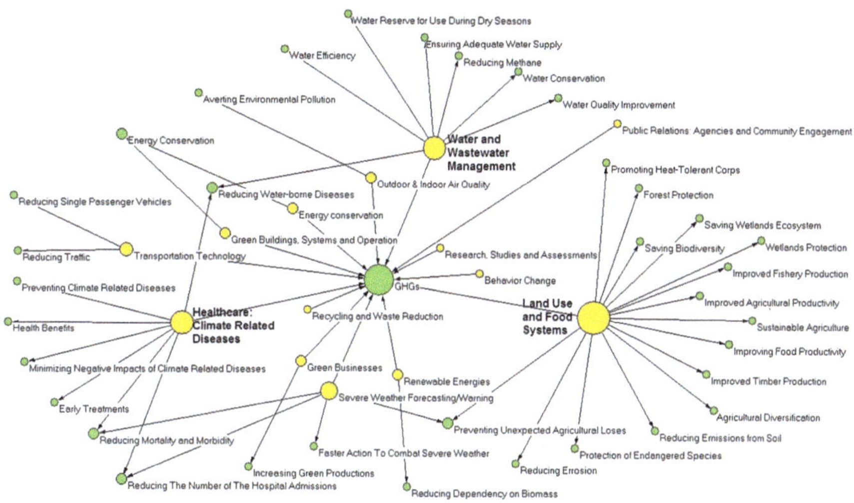

Fig. 7 Network diagram for the African city co-benefits from GHG emission reductions

economic co-benefits and co-costs. This is because economic and climate goals are still largely perceived as separate from each other and because city-level data to make predictions about job creation, income generation, diversification, and competitiveness is often lacking.

The network analysis for the US cities shows that green buildings and operations, transportation technology and land use and food systems are perceived to have the highest level of co-benefits among the 14 broad action categories identified for these cities. While transportation technology and green buildings and operations have overlapping co-benefits in terms of energy and fuel conservation and GHG reduction, land use and food systems show high ancillary benefits in sectors like agriculture, forestry, fisheries, and energy.

Among selected Asian cities belonging to the developing countries, all, except Karachi, have identified co-benefits in their climate action plans. The action co-benefits network diagram of Fig. 6 shows that the categories transportation technology, green buildings and operations, and water and wastewater management are perceived to produce the largest number of co-benefits. These three most influential climate action categories also show overlapping consequences for energy conservation, largely because this includes several cities in China where there is strong focus on energy sector benefits from low-carbon strategies. As expected, the generation of health sector co-benefits from improvements in transportation technology plays an important role in Asian climate action plans, given concerns about ambient air quality.

For selected African cities, the three action categories with the largest number of co-benefits are land use and food systems, water and wastewater management, and healthcare and climate change related diseases. With the burgeoning population in some African cities and the worst climate disasters faced in recent years, the co-benefits identified may expedite future climate mitigation actions because many of these actions are directly related to reducing resource scarcity and improvement of human health.

5 Insights and Implications from the Network Analysis of City Action Plans

The ranking of climate action categories and network analysis conducted here points at three sets of lessons. First, there are notable similarities and differences in the way co-benefits are perceived. For instance, for both US and selected Asian cities, the identification of co-benefits in transportation technology and green buildings and operations indicate a thrust of actions on energy conservation and efficient energy use. This is not surprising, given the cost reduction achieved through energy conservation measures in the majority of densely populated US and Chinese cities. Economic co-benefits through job creation are featured in both US and selected Asian city action plans. Many of these ancillary benefits, however, are less developed and often absent in the plans of African cities. Expanding the knowledge base on place-specific low-carbon strategies and their co-benefits may stimulate similar strategies in other locations.

Second, the networks reveal wide-ranging co-benefits from land use and food systems and from water and wastewater management in US and selected Asian cities. Both these actions necessitate community involvement in implementation and have long-run impacts on food security, energy efficiency, and water conservation. Co-benefits from such actions enhance the opportunities to link environmental sustainability and social cohesion directly with climate change mitigation plans.

Community participation ensures that local actors and households play an important role in addressing climate change risks among low income populations. These add to the equity and sustainability goals of the action plans

Third, the magnitudes of co-benefits from the perspectives of health, transportation, and air quality have typically been assessed through benefit cost analysis and simulation studies on future emission scenarios. Yet, except for some Chinese cities, the assessment of co-benefits is still at a nascent stage in most cities of developing Asia, Africa, and Latin America. The network analysis expands on these methods to identify relationships among co-benefits of climate action and thus provide structured input into the social discourse on prioritization of strategies.

In short, low-carbon strategies for cities can be more than about carbon, but about a host of economic, social, and environmental challenges that need to be addressed as cities grow and develop. Structured assessment of co-benefits and their inclusion in decision-making may help address these challenges.

Summary

- Cities are developing their own, specific action plans targeted at cutting their greenhouse gas emissions, reducing the impacts of climate-induced risks, and implementing measures to enhance community-level resilience to counter these risks.
- Such plans would include the direct effects of reducing contributions by cities to global atmospheric GHG concentrations, *plus* the co-benefits that are generated in terms of energy savings, reduced air pollution, and improved public health.
- In the absence of actual estimates of the magnitudes of co-benefits, a network diagram identifies the number of ancillary benefits from action plans targeted at lowering carbon emissions.
- For both US and Asian cities, the identification of co-benefits in transportation technology and green buildings and operations indicate a thrust of actions on energy conservation and efficient energy use.
- The networks reveal wide-ranging co-benefits from land use and food systems and from water and wastewater management in US and some selected Asian cities. Both these actions necessitate community involvement in implementation and have long-run impacts on food security, energy efficiency, and water conservation.
- Low-carbon strategies for cities can be more than about carbon, but about a host of economic, social, and environmental challenges. Structured assessment of co-benefits and their inclusion in decision-making may help address these challenges.

References

1. Working Group I: Contribution to the Fifth Assessment Report of the Intergovernmental Panel on Climate Change. Climate Change 2013: The Physical Science Basis, Summary for Policymakers. IPCC, Geneva, Switzerland (2014)
2. Ruth, M., World Energy Outlook: Energy Use in Cities, International Energy Agency. IEA, Paris, France (2008)
3. Rosenzweig, C., Solecki, W.D., Hammer, S.A., Mehrotra, S. (eds.): Urban Climate Change in Context. Climate Change and Cities: First Assessment Report of the Urban Climate Change Research Network, pp. 3–11. Cambridge University Press, Cambridge, UK (2011)
4. Ruth, M., Coelho, D.: Understanding and managing the complexity of urban systems under climate change. Clim. Policy. **7**, 317–336 (2007)
5. Worldwatch Institute: State of the World: Our Urban Future. W.W. Norton, New York (2007)
6. Aunan, K., Fang, J., Vennemo, H., Ove, K., Seip, H.M.: Co-benefits of climate policy—lessons learned from a study in Shanxi, China. Energy Policy. **32**, 567–581 (2004)

7. Aunan, K., Fang, J.H., Hu, T., Seip, H.M., Vennemo, H.: Climate change and air quality-measures with co-benefits in China. Environ. Sci. Technol. **40**, 4822–4829 (2006)
8. Chen, C.H., Wang, B.Y., Fu, Q.Y., Grenn, C., Streets, D.G.: Reductions in emissions of local air pollutants and co-benefits of Chinese energy policy: a Shanghai case study. Energy Policy. **34**, 754–762 (2006)
9. Harlan, S.L., Rudell, D.M.: Climate change and health in cities: impacts of heat and air pollution and potential co-benefits from mitigation and adaptation. Curr. Opin. Environ. Sustain. **3**, 126–134 (2011)
10. Burtraw, D., Krupnick, A., Palmer, K., Paul, A., Toman, M., Bloyd C.: Ancillary benefits of reduced air pollution in the united states from moderate greenhouse gas mitigation policies in the electricity sector. December 2001. Discussion paper 01–61, Resources for the future
11. Ng, Y.-K.: Welfare Economics: Introduction and Development of Basic Concepts, revised edn. Macmillan, Basingstoke, Hampshire (1983)
12. Pearce, D.: Policy frameworks for the ancillary benefits of climate change policies. In: OECD (ed.) Assessing the Ancillary Benefits and Costs of Greenhouse Gas Mitigation, pp. 517–560. OECD, Paris (2000)
13. Ekins, P.: How large a carbon tax is justified by the secondary benefits of CO_2 abatement? Resour. Energy Econ. **18**, 161–187 (1996)
14. McKinsey and Company: Pathways to a Low-Carbon Economy: Version 2 of the Global Greenhouse Gas Abatement Cost Curve (2009).
15. Dias, L.C., Domingues, A.R.: On multi-criteria sustainability assessment: spider-gram surface and dependence biases. Appl. Energy. **113**, 159–163 (2014)

Optimizing Water-Energy-Carbon Nexus in Cities for Low Carbon Development

Shobhakar Dhakal and Ashish Shrestha

Abstract Cities are the major consumer of water and energy, among other materials or resources. Carbon footprints of cities, especially in developing countries, are much higher compared to peri-urban and rural regions, with large contribution to national emissions. In the changing environment driven by increasing urbanization, with climate change and technological advancements, cities need to plan holistic as well as sectoral strategies to reduce emissions. Water and energy are two important sectors, which are often addressed as if they were separate from each other. While there is growing evidences and knowledge of their inherent interrelations, there are plenty of opportunities, which, if explored properly, can optimize water, energy, carbon footprints and contribute in low carbon development of cities while safeguarding future water and energy availability.

Keywords Water-energy-carbon nexus • Cities • Low carbon development

1 Introduction

Low carbon cities need to optimize many low carbon opportunities in the urban systems across all sectors. Water and energy are two important urban sectors, which are directly or indirectly linked to all other urban systems. Optimization of water and energy resources will secure the future demands with minimal implications of environmental damage and greenhouse gas (GHG) emissions. Water and energy are inherently linked, with many common drivers such as population growth, climate change, urbanization, and affluence, which are increasing consumption. Water is required for energy processing, production, and supply; energy is required for water

S. Dhakal (✉)
School of Environment, Resources and Development and Department of Energy, Environment and Climate Change, Asian Institute of Technology, Pahaolyothin Road, Klong Luang, Pathumthani 12120, Thailand
e-mail: shobhakar@ait.ac.th; shobhakar.dhakal@gmail.com

A. Shrestha
Department of Energy, Environment and Climate Change, Asian Institute of Technology (AIT), Klong Luang,
Pathumthani 12120, Thailand

S. Dhakal, M. Ruth (eds.), *Creating Low Carbon Cities*,
DOI 10.1007/978-3-319-49730-3_4

transport, treatment, and supply, creating the nexus between the demand and resource utilization of water and energy. Primary energy extraction, refining, processing, and transport require huge amounts of water, and mostly freshwater is used. For example, production of crude oil requires approximately 40 l of water per l of oil, non-conventional oil requires 90–150 l of water per l of oil, coal requires 4 l of water per kg of coal, natural gas requires 6 l of water for a kg of gas produced, and biomass requires 1100 l of water per l of ethanol produced [1, 2].

Cities are complex systems, where energy and water are major inputs and wastes and emissions as outputs. In many cities, most of the energy production is located outside the city boundary and urban water infrastructures are located inside the city boundary. This chapter mostly focuses on issues related to energy for water in cities context. Urban Water Systems (UWS), including municipal drinking water supply and waste water management systems, have a sizeable share in the total energy use and GHGs emission of cities. In the context of expanding urbanization, the energy footprints of the UWS will continue to grow, resulting also in increases in overall urban water related carbon footprints.

In the broader global picture, water demand is projected to increase by about 55 % by 2050 to meet growing demands from industrial sectors, thermal electricity generations, and domestic uses [3]. Much of this increase is attributable to a near-doubling in the world's urban population by 2050 to 6.4 billion from the current (2014) 3.9 billion [4]. At the same time, about 11.1 % of global population is without access to improved water and 18.1 % of global population is without access to electricity in 2011 [5]. Many research findings unambiguously agree that cities are the major contributor to climate change, and they are also most affected by the impacts of climate change [6, 7]. Despite the fact that they cover less than 2 % of the earth's surface, cities consume 78 % of the world's energy and produce over 71 % of global energy related CO_2 emissions and more than 60 % of total GHG emissions combining energy generation, vehicles, industry, and biomass use [8].

The UWS is one of the key sector which must be explored to optimize urban energy and carbon footprints as UWS are reported to utilize 1–18 % of total energy use in the city depending on the local context of cities [1]. Generally, there are a number of driving forces which affect water-energy-carbon interrelation, including technologies of the treatment systems; size, capacity, and utilization of civil and electromechanical components used in the water/wastewater treatment, transport, distribution, and collection units; water and wastewater quality standards; design of the drinking water distribution systems; age of the infrastructures; climate change and its implications on UWS; and other environmental regulations.

The UWS is complex and diverse, involving processes of water services delivery to different economic sectors, and which rely on energy throughout each element of the UWS. Typically, fossil fuels are primary sources of energy, which produce considerable amounts of GHGs. Water security and energy security have risen to become focal issues during the city's growth and transformation and significant resources have been spent in both the developed and developing countries to build UWS infrastructures.

2 Characterizing Urban Water Systems

Cities today have massive infrastructures to deliver ever-increasing water demand and collect wastewater and storm water for safe removal of used and polluted water. Cities' local governments take responsibility for master planning, execution, operation, and maintenance of these infrastructures. Globally, Asia withdraws the highest water volume from surface and ground water resources (2378 Gm^3/year), followed by North America (525 Gm^3/year), Europe (418 Gm^3/year), Latin America (252 Gm^3/year), Africa (217 Gm^3/year), Ocenia (26 Gm^3/year), and Caribbean (13 Gm^3/year) [9].

UWS consists of a series of processes (Fig. 1) from abstraction of water from surface or ground resources, transport or conveyance of raw water in bulk through single or multiple channels to treatment plants, a series of treatment processes, distribution of treated water to households and industries, and ultimate use of water or consumption for different purposes. Subsequently, the management of wastewater occurs through collection of used water (usually along with storm water) through open or closed collection systems or conduits to wastewater treatment plants and final disposal of treated wastewater to water bodies complying to environmental regulations and standards. Groundwater aquifers are diminishing, it was estimated 20 % of the world's aquifers are being overexploited [5] with the rate of abstraction increasing by 1–2% per year [10]. With ground water depletion being growing issue, surface water availability becoming limited and transport of water over long distance is very costly, desalination has become a feasible alternative in most of the developed coastal cities which are very energy and carbon intensive. At the municipal level, conventional water treatment systems of UWS consist of primary, secondary, and tertiary treatment. There is considerable development in micro filtration or membrane filtration technology. Despite being energy intensive, these filtration technologies have applications both at the municipal and at the household level. In developed country cities, almost all settlements are served by water distribution networks. However, in developing country cities partial supply to the unserved community are met by delivering water through tankers (water tankering). Water distribution system consists of massive structure for supply mains, arteries, structures with maximum capacitics for firefighting and number of pumps to maintain pressure within the networks, which is recommended to be 450–520 kPa for common practices for buildings up to ten stories. The minimum pressure in the water distribution main should be 280 kPa for residential service connection [11]. All of these need energy and have carbon implications.

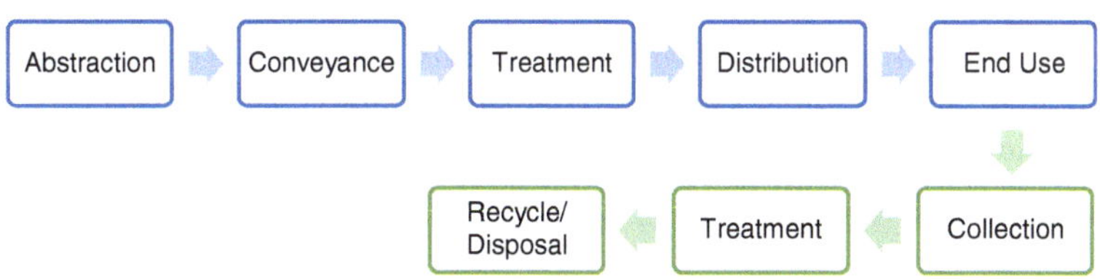

Fig. 1 Different elements of urban water system

Many cities and communities are reusing wastewater for nonconsumptive purposes, e.g., grey water from showers and sinks in households for gardening, car washing, and toilet flushing through minimal treatment or filtration. For an example, Singapore is extensively using local catchment water or rain water as one of the major source of drinking water. Two-third of Singapore erves as water catchment to harvest rainwater through 17 reservoirs. Singapore is also reusing wastewater which is known as "NEWater," through advanced membrane technologies, micro filtration and reverse osmosis, which currently supply 30 % of the country's need and which is being planned for expansion to meet 55 % water demand by 2060 [12]. The end use of UWS too is one of the crucial domain which has significant impact on water related energy use in cities. Use of individual household level water purification systems, use of hot water for drinking and cleaning, as well as use of water for food production, processing, and preparation contributes with significant footprints of energy and carbon. Furthermore, use of bottled water in cities contributes to significant carbon footprint's share of water resource consumption in cities which is mainly attributable to bottling itself and shipment through fossil fuels based vehicular transports.

3 Energy Dependency of Urban Water Systems

The energy required for the various processes of the UWS varies across cities and depends on numbers of local parameters such as climatic conditions, source of water, water availability, population and affluence, treatment standards, topography, related policies, and more. Energy is usually used in electromechanical components. The total energy consumption in UWS can be categorized into following aspects:

- Net energy use in UWS (energy consumed in all stages of UWS) is the sum of all the direct operational energy uses and the sum of all the embodied energies of materials. Energy recovery in term of biogas and chemicals are a sustainable option adopted by comparatively fewer utilities that contribute to negative energy and carbon footprints. Net energy use of UWS can be described as following relation:
  Net energy use of UWS = Operational energy in UWS + Embodied energy of UWS infrastructures − Energy recovery from treatment by−products (mainly after wastewater treatment in UWS)
- Operational energy of UWS can be defined as all the direct energy uses like electricity and fuels required within all stages of UWS, which can be represented in term of intensity of electricity consumed per unit volume of water, i.e.,
  Operational Energy of UWS (kWh/m^3) = Sum of electrical energy consumption in all elements of UWS divided by respective amount of water processed/

treated (kWh/m^3) + other fuel sources for transport of water or wastewater including sludge transports divided by respective amount of water transported/ processed (kWh/m^3)

- Embodied energy is the sum of all energy required for the production of construction materials and chemicals in UWS. Most of the studies on embodied energy of UWS are based on Life Cycle Assessment (LCA), a systematic and quantitative evaluating tool for the impacts of materials, products, processes, or services from "cradle" to "grave." Some examples of relevant LCA papers are for Chengdu, China [13], Los Angeles, Southern California [14], and Sydney, Australia [15], Canada [16]. Other comprehensive approaches for estimating embodied energy include process-based hybrid approaches and input-output based hybrid approaches [17]. Corominas et al. [18] provide a comprehensive review of research on wastewater treatment and applications of LCA.

Wastewater Energy requirement to provide safe drinking water for human consumption depends on types of water sources. In general, lake or river water requires 0.37 kWh/m^3, groundwater requires 0.48 kWh/m^3, wastewater reuse requires 1.0–2.5 kWh/m^3 and seawater desalination requires 2.58–8.5 kWh/m ^{3} [19]. Groundwater is the primary source of drinking water in many countries (Denmark 99 %, Mexico 95 %) while it is 38 % in USA [5]. In term of water volume, India (0.251 million m^3/year), China (0.112 million m^3/year), and the USA (0.112 million m^3/year) are top water abstracting countries in year 2010 [20]. While in other countries groundwater abstraction have been regulated and prohibited due to environmental issues including depleting ground water and ground subsidence. While groundwater, without localized contamination and salt water intrusion, in general is of higher quality than surface water, with lesser microbial contamination, groundwater pumping typically requires around 0.1 kWh/m^3 at 36.5 m depth to 0.5 kWh/m^3 at 122 m depth [21]. More stringent water quality standards and new disinfection technologies such as ozone and ultraviolet disinfection (UV) will likely increase energy demand in the water sector. High service pumps in distribution networks, raw water pumping, treatment units, controls, and lighting are the largest energy consumption process in UWS. Energy required to deliver water varies greatly between water utilities influenced by geography, topography, climate, and available infrastructures. The source and water quality have a significant impact on overall energy uses in UWS. For wastewater treatment process, energy requirements are influenced mostly by treatment standards and regulations. Primary and secondary treatment processes in both water and wastewater consumes less energy compared to their tertiary treatments. A range and average figures for energy footprints in water and wastewater utilities based on various references on different countries or cities are summarized in Table 1.

Apart from municipal UWS, many people also intensively use polyethylene terephthalate (PET) bottled water. Global bottled water consumption has increased from around 100,000 million liters in 1998 to around 224,000 million liters in 2008 and was projected to be over 250,000 million liters by 2013 [30]. As convenient and beneficial the use of bottled water is, it has environmental implications as water,

Table 1 Energy footprints of water service delivery in different countries

Countries	Energy requirements	Energy intensity (kWh/m^3)		References
		Range	Average	
Australia	Energy: water utilities	0.09–1.84	0.72	Kenway et al. [22]
	Energy wastewater utilities	0.47–1.13	0.77	
United States	Production and distribution of potable water in Western US	1.32–3.96	–	Valentina et al. [23], Wilkinson [24], U.S. Department of Energy [25]
	Production and distribution of potable water in Eastern US	0.48–0.66	–	
	Range for water supply utilities	0.08–1.00		Carlson and Walburger [26]
	Range for wastewater utilities	0.20–0.90		
	California: water conveyance	0.00–1.06	–	Valentina et al. [23]
	California: water Treatment	0.03–4.23	–	
	California: water Distribution	0.18–0.32	–	
	California: wastewater collection and treatment	0.29–1.22	–	
Germany	Water conveyance and treatment	0.12–1.13	–	Valentina et al. [23]
	Water distribution	0.03–0.58	–	
	Wastewater collection and treatment	0.39–0.83	–	
Singapore	NEWater for uses such as industry	0.7–1.2	0.95	Lenouvel et al. [27]
	Seawater desalination	3.9–4.3	4.1	
	Wastewater treatment	0.52–0.89		
Norway (Oslo)	Electricity use in water treatment and supply (2000–2006)	0.38–0.44	0.40	Venkatesh and Brattebo [28]
	Electricity use in wastewater collection and treatment (2000–2006)	0.67–0.87	0.80	
Thailand (Bangkok)	Water supply utilities	–	1.59	Dhakal et al. [29]
	Wastewater utilities	–	2.16	

energy, and carbon footprints of bottled water is very high. The water footprint of bottled water constitutes the real water content and virtual water use for all the water production processes, including water used for manufacturing of raw materials. A study in India [31] showed that one liter of bottled water requires 17.41 l of water and 7.08 MJ (1.97 kWh) of energy to make it available. The average figure for North America bottled water showed 1.32 l of water and 0.24 MJ (0.067 kWh) of energy use per liter of bottled water produced [32]. The real water consumption behind bottled water in Italy showed that it required average of 2.29 l of water to supply 1 l of bottled PET water to users [33].

Wastewater treatment is also an energy intensive process because it involves several treatment processes to remove pollutants for safe disposal to water bodies. Several technologies for treatment have different energy footprints. Membrane bioreactors are more energy intensive than conventional activated sludge. In wastewater treatment process, the most energy consuming processes are aeration for biological treatment, and pumping, mechanical treatment, and ventilation for odor control. The typical breakdown of energy consumption for French conventional wastewater treatment plants with nutrient removal showed that maximum energy use of the total energy used in WWTP are in aeration of activated sludge (45 %), auxiliary equipment and pumping (18 %), odor treatment (12 %), pre-treatment, primary settling, and biological tank mixing (11 %), and sludge dewatering, digestion, thickening, and recirculation consumed 14 % [23]. From several databases from the United States, of all the processes in UWS, water treatment is the highest energy intensive process in drinking water sector, which consumes energy approximately from 0.3 to 1.5 kWh/m^3 on average, while aerated lagoon and oxidation ditch consumes more energy in wastewater sector at the range of 0.5–5.9 kWh/m^3 and 0.9–4.9 kWh/m^3, respectively [23].

4 Drivers That Affect Water Related Energy Linkage in Cities

It is important to understand drivers that shape the WEC nexus, to implement any policies, design future systems, or plan strategies to mitigate GHGs emissions. Some of the important drivers are briefly highlighted below.

- **Climate change**—Various implications of climate change are observed in water sectors; one such example is the issue of water availability due to frequency of extreme events such as floods, droughts, and heat waves. Climate change creates uncertainty about trends and extremes of future climate variables. The normal climate trend is altered with the effects of climate change, as a result, some basin receives excess water while other basins suffers drought amplifying the already intense competition for water resources. Groundwater table and river discharges, for example, are affected hence creating pressure on water services management. In cities, heat island phenomenon also affects the consumption of water and energy. Due to the urban heat island effect, temperature difference in some cities between urban core and surrounding rural areas could reach to difference of 10°, in US cities (with over 1,000,000 people) for every 1 °C increase in temperature, peak electricity loads increase by about 1 %. For example, in Los Angeles, there is net rate of increase of 167 MW per 1 °C increment. Also, in Toronto, a 1 °C increase on summer days links to 1.6 % increase in peak electricity demand [34]. The added strain on the energy generation and distribution system consequently result in additional water consumption for energy generation processes. Climate change is also one of the contributing factors that drives change in technology and infrastructure in water utilities.
- **Increase in population**—From 1800 to 1930, the world's population doubled from one billion to two billion, and it doubled from three billion to six billion just over from 40 years i.e., 1960–1999. Today the world's population exceeds 7.4

billion. UN population division projects that the population will surpass nine billion by 2050. Population growth is not only directly related to increased water and food demand, the consequences are linked with many indirect impacts including contribution to GHG emissions and use of high quantity of fossil fuels.

- **Increase in urbanization**—Urban populations are growing because of an overall rise in population as well as migration. Today the Asia's urban population is 44 %, which is expected to reach 64 % at the middle of the century [35]. For the developing world, it is expected that by 2030, 56 % of the population will live in cities. The major challenge involved with urban areas is the unpredictability and migration trend, in order to ensure proper water and energy services.
- **Change in technologies**—Advanced technologies are being implemented in water utilities which are generally energy intensive. Technology such as membrane based reverse osmosis consumes higher amount of energy compared to conventional system using coagulation and flocculation and rapid sand filtrations. Most of the cities are now conveying water from inter-basin sources over long distances. The decrease in fresh water availability is also one of the factors for shifting towards alternate source such as desalination. Different desalination techniques involve result in different energy uses [36]:

  - Single stage evaporation: 650 kWh/m^3
  - Multistage flash: 55–80 kWh/m^3
  - Multi-effect distillation: 40–65 kWh/m^3, and
  - Reverse osmosis: 3.7 kWh/m^3

One particular study in California showed that if water supply from desalination is implemented, the electricity consumption would be 52 % of total electricity used in the state [37].

- **Aging infrastructures**—Asset management in water utilities has been a growing priority in research and development. Ageing infrastructures have consequences for water leakage at cost of both water and energy. The water pricing mechanism plays major roles in controlling water losses. Water losses in some developing countries exceed 40 %. Prevention of physical water loss means reducing energy and carbon footprints in water services.
- **Regulations**
  Pollution—The wastewater generated through the water consumption process is treated up to the safe disposal threshold limit for each and every pollutant. Regulations are often maintained by city authorities normally under guidelines of World Health Organization (WHO). Regulation is very important in order to prevent water sources contamination, soil salinization, ground water pollution, and health related hazards. Furthermore, pollution charges for discharging wastewater to water bodies can drive innovation in recycling or additional treatment.
  Environmental Consideration—Environmental flows during water abstraction and its associated impacts on ecosystems are well addressed in some nations while in others it is still neglected. Research in Asia and the Pacific showed that 23 out of 48 countries are undertaking activities to integrate environmental flows into local, regional, and state level planning processes [38].

5 Towards Net Zero GHG Emission and Self-Sufficiency

The net energy use of the UWS will increase under the influence of increasing urbanization, climate change implications, and increasing water demand. The shift towards cleaner energy sources in the full scale will not happen soon because of financial constraints and inertia associated with the replacement of (recently) developed infrastructures. On a positive note there is significant realization of the need for a clean energy paradigm. The United Nations Secretary General's "Sustainable Energy for All" initiative is driving the expansion of renewables in the aggregated global energy mix. Several countries, as part of this agenda, set ambitious targets to double the share of renewables in the mix by 2030. European Union also committed to reduce overall GHG emissions by 40 % by 2030 compared to 1990 level as well as to increase renewable energy share at least to 27 %. Furthermore, Energy Roadmap 2050, developed by European Commission, aims to reduce greenhouse gas emissions by 2050 by over 80 %. Meeting ever-growing energy demands requires coherence between water use and climate change mitigation. UWS has the significant role in mitigating GHGs and support cities emission reduction targets. Renewable energy sources like wind and solar photovoltaic (PV) power have very low carbon footprints and lower water consumption per unit of energy delivered than carbon-based fuels.

There are plenty of opportunities in UWS where water/wastewater utilities can recover energy and use decentralized renewable sources. Some measures are improving their existing systems such that its efficiency are at the optimum scale all the time and some measures require distinct addition to their systems such as biogas recovery, solar and wind power generation as well as redesigning the system when possible.

In most developing cities, non-revenue water (NRW) loss is still very high. In Asian countries NRW levels ranged from 5 to 56 % in 2009 [39]. Reduction in real water losses eases burden on the supply side, and water security improves on the demand side. Both have augmenting beneficial implications, as energy uses in UWS and GHG emissions abate.

The future of water and energy consumption patterns in new and expanding cities can be optimized during the urban planning process. Compact settlements have lower footprints of water distribution and wastewater collection infrastructures. In designing new systems, embodied energy shall be considered and should be optimized where possible. Operational energy depends on type of systems: decentralized versus centralized, scales of UWS utilities and their capacity utilization. The optimum operating conditions have minimum water, energy, and carbon footprint.

The direct and indirect reuse of water can contribute in reducing WEC footprints. Reuse of treated wastewater to potable standards might be beneficial in some contexts where natural water sources are located too far from consumers or when water sources are overexploited. Reuse of wastewater for potable water generally might have higher energy footprints. However, different scenarios should be considered in the planning and design phase. There are many practical applications of water reuse for non-potable uses such as using grey water recycling for toilet flushing or gardening. Rainwater harvesting at some location eases water security at the household level as well as reduces stress on water utilities. Reuse of rainwater or storm water for non-potable uses require less amount of energy compared to reusing wastewater due to more contaminants removal required in wastewater.

Wastewater and by-products of treatment processes are resources in themselves as they contain energy and nutrients. The energy in the wastewater includes potential energy, thermal energy, and chemical energy. It is the matter of design and feasibility to extract these energy forms. For transitioning to net zero GHG emission, such possibilities shall be engineered accordingly. The concept of kinetic energy recovery from wastewater is not common, but many possible applications have been studied through generation of hydropower using small turbines [40, 41]. There are increasing research and development interests on multi-purpose hydropower schemes that fit turbines wherever gravity flow is possible, to existing drinking water transport conduits the wastewater network, navigation locks and dams, desalination plants and heating and cooling systems. The most suitable locations for hydropower generation are break pressure tanks, pressure reducing valves, and municipal water supply dams. Furthermore, other potential sites include outlets and inlets of WWTPs, especially the larger capacity plants with high flow rates [40]. Thermal energy recovery from wastewater is another possibility that provides renewable energy, mostly applicable in buildings and sewer systems, using heat exchangers and heat pumps. Several studies reveal the applicability of thermal energy recovery from wastewater [42, 43]. Chemical energy recovery using anaerobic digestion is more commonly practiced but largely limited to developed cities. The extraction of biogas replaces fossil fuels, reduces the amount of sludge to be disposed which also have financial benefit to utilities operation. With proper planning and design, UWS can achieve net zero emission as well as become self-

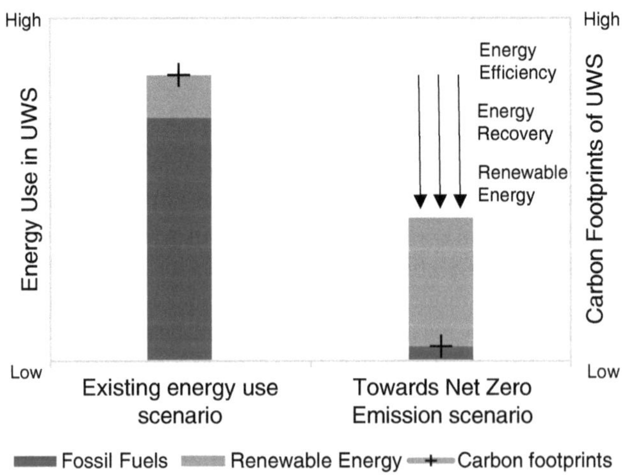

Fig. 2 Towards scenario of net zero emission in UWS

sufficient in term of energy use. Using technologies such as co-generation by addition of organic waste to the digesters, utilizing thermal energy of the wastewater for space heating and alternative wastewater and waste options using alternative processes, municipal wastewater systems can even become "energy-positive." As an example, a study in Austria showed wastewater treatment plants are capable of reaching up to 180 % energy generation compared to the energy needs [44].

There are significant opportunities to transition from current fossil fuel intensive UWS to new scenarios with net zero emissions (Fig. 2). Supplying the demand with high quality assurance, energy efficiency shall be improved along with extensive recovery of energy and nutrients in UWS processes and shift towards self-sufficiency by relying on renewable energy sources.

6 Conclusion

As there is a growing need for cities' transition into a cleaner, healthier, sustainable, and economically secured future, there are a number of approaches that cities must adopt in water-energy systems, including investments in renewable technologies, improving efficiency of water and energy systems, reforming the necessary regulations and policies. Cities play a significant role in determining the future of water and energy resources as well as combating climate change. Cities have significant impact on the upstream of water and energy cycles. Water and energy planners need to understand the drivers that influence the water-energy nexus at the city scale and beyond while formulating policies that help in minimizing their energy-carbon footprint. Planning of water and energy sectors requires inter-sectoral and organizational coordination at all governing levels. Moreover, water and energy are two key areas for sustainable development and therefore addressing them simultaneously, understanding their dependencies, synergies, conflicts, and trade-offs are vital before planning any development strategies. This would enhance mutually beneficial aspects and mitigate any adverse implications through either sectors. Therefore, there is a need for proper governance and institutional arrangements, as it is the prerequisite to plan optimization strategies for the water, energy, and carbon nexus. There is also a need for more comprehensive research to assess the individual as well as integrated components of WEC nexus, and develop necessary guidelines and management tools to support decision makers.

References

1. Olsson, G.: Water and Energy: Threats and Opportunities. IWA Publishing, London (2012)
2. World Energy Council: Water for Energy 2010. World Energy Council, London, UK (2010) ISBN: 978-0-946121-10-6. http://www.worldenergy.org/wp-content/uploads/2012/10/PUB_Water_For_Energy_2010_WEC.pdf

3. UN Water: Water for sustainable world Executive Summary. The United Nations World Water Development Report 2015. United Nations World Water Assessment Programme. http://unesdoc.unesco.org/images/0023/002322/232272E.pdf (2015). Accessed 20 July 2016
4. United Nations, Department of Economic and Social Affairs, Population Division (2014) World Urbanization Prospects: The 2014 Revision, Highlights (ST/ESA/SER.A/352) (2014)
5. World Water Assessment Programme (WWAP): The United Nations World Water Development Report 2014: Water and Energy. UNESCO, Paris (2014)
6. Revi, A., Satterthwaite, D.E., Aragón-Durand, F., Corfee-Morlot, J., Kiunsi, R.B.R., Pelling, M., Roberts, D.C., Solecki, W.: Urban areas. In: Field, C.B., Barros, V.R., Dokken, D.J., Mach, K.J., Mastrandrea, M.D., Bilir, T.E., Chatterjee, M., Ebi, K.L., Estrada, Y.O., Genova, R.C., Girma, B., Kissel, E.S., Levy, A.N., MacCracken, S., Mastrandrea, P.R., White, L.L. (eds.) Climate Change 2014: Impacts, Adaptation, and Vulnerability Part A: Global and Sectoral Aspects Contribution of Working Group II to the Fifth Assessment Report of the Intergovernmental Panel on Climate Change, pp. 535–612. Cambridge University Press, Cambridge, UK (2014)
7. Seto, K.C., Dhakal, S., Bigio, A., Blanco, H., Delgado, G.C., Dewar, D., Huang, L., Inaba, A., Kansal, A., Lwasa, S., McMahon, J.E., Müller, D.B., Murakami, J., Nagendra, H., Ramaswami, A.: Human settlements, infrastructure and spatial planning. In: Edenhofer, O., Pichs-Madruga, R., Sokona, Y., Farahani, E., Kadner, S., Seyboth, K., Adler, A., Baum, I., Brunner, S., Eickemeier, P., Kriemann, B., Savolainen, J., Schlömer, S., von Stechow, C., Zwickel, T., Minx, J.C. (eds.) Climate Change 2014: Mitigation of Climate Change. Contribution of Working Group III to the Fifth Assessment Report of the Intergovernmental Panel on Climate Change. Cambridge University Press, Cambridge, UK (2014)
8. UN Habitat: Climate Change. http://unhabitat.org/urban-themes/climate-change/ (2012). Accessed 20 July 2016
9. Plappally, A.K., Lienhard, V.J.H.: Energy requirements for water production, treatment, end use, reclamation, and disposal. Renew. Sustain. Energy Rev. **16**, 4818–4848 (2012)
10. World Water Assessment Programme (WWAP): United Nations world water development report 4: Managing water under uncertainty and risk. UNESCO, Paris (2012)
11. Jiri, M., Blanca, J.C., Mohammad, K., Per-Arne, M., Joel, A.G., Bernard, C.: Urban Water Cycle Processes and Interactions: Urban Water Series—UNESCO-IHP. CRC Press, Boca Raton (2008)
12. PUB (Singapore's National Water Agency): Four National Taps. https://www.pub.gov.sg/watersupply/fournationaltaps (2016). Accessed 25 Aug 2016
13. Murray, A., Horvath, A., Nelson, K.L.: Hybrid Life-cycle environmental and cost inventory of sewage sludge treatment and end-use scenarios: a case study from China. Environ. Sci. Technol. **42**, 3163–3169 (2008)
14. Fang, A.J., Newell, J.P., Cousins, J.J.: The energy and emissions footprint of water supply for Southern California. Environ. Res. Lett. **10**, 114002 (2015). doi:10.1088/1748-9326/10/11/114002
15. Lundie, S., Peters, G.M., Beavis, P.C.: Life cycle assessment for sustainable metropolitan water systems planning. Environ. Sci. Technol. **38**, 3465–3473 (2004)
16. Sahely, H.R., MacLean, H.L., Monteith, H.D., Bagley, D.M.: Comparison of on-site and upstream greenhouse gas emissions from Canadian municipal wastewater treatment facilities. J. Environ. Eng. Sci. **5**, 405–415 (2006)
17. Weiwei, M., Qiong, Z., James, R.M., David, R.H.: Embodied Energy Comparison of Surface Water and Groundwater Supply Options. Water Res. **45**, 5577–5586 (2011)
18. Corominas, L.I., Foley, J., Guest, J.S., Hospido, A., Larsen, H.F., Morera, S., Shaw, A.: Life cycle assessment applied to wastewater treatment: state of the art. Water Res. **47**, 5480–5492 (2013)
19. World Business Council on Sustainable Development (WBCSD): Water, Energy and Climate Change: A Contribution from the Business Community. WBSCD, Geneva (2009)

20. United Nations Educational, Scientific and Cultural Organization (UNESCO): Groundwater and global change: trends, opportunities and challenges. United Nations World Water Assessment Programme. http://unesdoc.unesco.org/images/0021/002154/215496e.pdf (2012). ISBN 978-92-3-001049-2
21. United States Government Accountability Office (US GAO): Energy-water nexus: amount of energy needed to supply, use and treat water is location-specific and can be reduced by certain technologies and approaches. Report to the Ranking Member Committee on Science, Space and Technology, US House of Representatives. GAO-11-225. US GAO, Washington, DC (2011)
22. Kenway, S.J., Priestley, A., Cook, S., Seo, S., Inman, M., Gregory, A., Hall, M.: Energy use in the Provision and Consumption of Urban Water in Australia and New Zealand. CSIRO: Water for a Healthy Country National Research Flagship, CSIRO Australia and Water Services Association of Australia, Canberra, Australia (2008)
23. Valentina, L., Kwang-Ho, C., Peter, C.: Water-energy interactions in water reuse. IWA Publishing, London (2012)
24. Wilkinson R: Methodology for analysis of the energy intensity of California's water systems, and an assessment of multiple potential benefits through integrated water-energy efficiency measures. Environmental Sciences. California Institute for Energy Efficiency, January 2000, p. 89 (2000)
25. U.S. Department of Energy: Energy demands on water resources. Report to Congress on the Interdependency of Energy and Water, December 2006 (2006)
26. Carlson, S.W., Walburger, A.: Energy Index development for benchmarking water and wastewater utilities. AWWA Research Foundation, California Energy Commission, and New York State Energy Research and Development Authority, Denver, CO (2007)
27. Lenouvel, V., Lafforgue, M., Chevauché, C., Rhétoré, P.: The energy cost of water independence: the case of Singapore. Water Sci. Technol. **70**(5), 787–794 (2014). doi:10.2166/wst.2014.290
28. Venkatesh, G., Brattebø, H.: Energy consumption, costs and environmental impacts for urban water cycle services: case study of Oslo (Norway). Energy. **36**(2), 792–800 (2011)
29. Dhakal, S., Shrestha, S., Shrestha, A., Kancko, S., Kansal, A.: Understanding and quantifying the water-energy-carbon nexus for low carbon development in Asian cities. Kobe: Asia-Pacific Network for Global Change Research (under publication). (2016)
30. Nicholas, D.: Technology of Bottled Water, 3rd edn. Wiley-Blackwell, UK (2011)
31. Shalini, A.T., Niranjan, K., Rakesh, K.: Water and energy footprint assessment of bottled water industries in India. Nat. Resour. **5**, 68–72 (2014)
32. International Bottled Water Association (IBWA): Water and energy use benchmarking study: executive summary. http://www.bottledwater.org/education/environment (2014). Accessed 25 July 2016
33. Niccolucci, V., Botto, S., Rugani, B., Nicolardi, V., Bastianoni, S., Gaggi, C.: The real water consumption behind drinking water: the case of Italy. J. Environ. Manage. **92**(10), 2611–2608 (2011). doi:10.1016/j.jenvman.2011.05.033
34. Kennedy, C., Cuddihy, J., Engel-Yan, J.: The changing metabolism of cities. J. Ind. Ecol. **11**(2), 43–59 (2007)
35. United Nations, Department of Economic and Social Affairs, Population Division: World Urbanization Prospects, the 2011 Revision, New York (2012)
36. Semiat, R.: Energy issues in desalination processes. Environ. Sci. Technol. **42**(22), 8193–8201 (2008)
37. Stokes, J.R., Horvath, A.: Energy and air emission effects of water supply. Environ. Sci. Technol. **43**(8), 2680–2687 (2009)
38. World Water Assessment Programme (WWAP): The United Nations World Water Development Report 3: Water in a Changing World. UNESCO, Paris (2009)
39. Asian Development Bank: The Issues and Challenges of Reducing Non-revenue Water, p. 2010. Asian Development Bank, Mandaluyong City, Philippines (2010)

40. Christine, P., McNabola, A., Paul, C.: Development of an evaluation method for hydropower energy recovery in wastewater treatment plants: case studies in Ireland and the UK. Sust. Energy Technol. Assess. **7**, 166–177 (2014)
41. Kyu, J.C., In-Soo, K., Xianghao, R., Kyeong-Ho, C.: Reliable energy recovery in an existing municipal wastewater treatment plant with a flow-variable micro-hydropower system. Energ. Conver. Manage. **101**, 681–688 (2015)
42. Florian, K., Lena, S., Thomas, E.: Analysing wastewater temperature development in a sewer system as a basis for the evaluation of wastewater heat recovery potentials. Energ. Buildings. **128**, 639–648 (2016)
43. Oguzhan, C., Huseyin, G., Emrah, B., Orhan, E., Arif, H.: Heat exchanger applications in wastewater source heat pumps for buildings: a key review. Energ. Buildings. **104**, 215–232 (2015)
44. Otto, N., Peter, E., Petar, V.: Ways to optimize the energy balance of municipal wastewater systems: lessons learned from Austrian applications. J. Clean. Prod. **88**, 125–131 (2015)

Dr. Shobhakar Dhakal is an associate professor in the Department of Energy, Environment and Climate Change at the Asian Institute of Technology. He received his Ph.D. from University of Tokyo in 2000. His teaching, research, and international activities comprise energy and climate policy, energy system analysis, accounting of energy and carbon emissions and mitigation opportunities, and low carbon cities. He was a Coordinating Lead Author of the Inter-Governmental Panel on Climate Change (IPCC) for its Fifth Mitigation Assessment report (released in April 2014) for the Chapter on Human Settlements, Infrastructure and Spatial Planning (2010–2015).
shobhakar@ait.ac.th, shobhakar.dhakal@gmail.com, School of Environment, Resources and Development, and Department of Energy, Environment and Climate Change, Asian Institute of Technology (AIT), Klong Luang, Pathumthani 12120, Thailand

Mr. Ashish Shrestha is a research associate in the Department of Energy, Environment and Climate Change at the Asian Institute of Technology involved in the research project on water-energy-carbon nexus in Asian cities. He received his dual degree M. Engg. in urban water engineering and management from Asian Institute of Technology and UNESCO-IHE Institute for Water Education in 2013, and B. Tech. in environmental engineering from Kathmandu University in 2009. His research areas and interests include climate change modeling, water-energy nexus, urban water systems, urban hydrology, and environmental engineering.
ashish@ait.ac.th, ashish.shrs@gmail.com, Energy Field, Asian Institute of Technology (AIT), Klong Luang, Pathumthani 12120, Thailand

Grassroots Environmentalism and Low-Carbon Cities

Kevin Lo

Abstract Grassroots environmentalism has emerged as a key driver for low-carbon cities. This chapter describes and analyses three cases of grassroots environmentalism: the transition town movement, the ecovillage movement, and community renewable energy initiatives. It shows that, rather than business-as-usual, advocates of grassroots environmentalism envision a fundamental transformation of everyday practices based on community building, resilience, economic localisation, and dematerialisation.

Keywords Grassroots environmentalism • Low-carbon cities • Climate change • Transition town • Ecovillages • Community renewable energy

1 Introduction

Increasingly, we are witnessing the emergence of grassroots environmentalism in the context of climate change mitigation and sustainable energy transition. This trend is underpinned by two factors. First, public awareness of climate change and its impacts has risen to unprecedented levels. Second, there is a growing perception that existing low-carbon responses are insufficient by themselves to prevent dangerous climate change [1]. Rather than despairing at the lack of progress, concerned citizens around the world choose to reduce carbon emissions individually as well as collectively in the forms of community action. Recent research suggests that emerging low-carbon community action is excitingly diverse, creative, ambitious, and transformative, being concerned with the issues of climate justice, resilience, and radical alternatives to conventional low-carbon approaches [2, 3]. This chapter examines the roles and limitations of grassroots environmentalism in shaping low-carbon cities using three prevalent cases as a prism: the transition town movement, the ecovillage movement, and community renewable energy initiatives.

K. Lo (✉)
Department of Geography, Hong Kong Baptist University, Kowloon Tong, Hong Kong
e-mail: lokevin@hkbu.edu.hk

S. Dhakal, M. Ruth (eds.), *Creating Low Carbon Cities*,
DOI 10.1007/978-3-319-49730-3_5

2 Transition Town Movement

The transition town movement was founded by Rob Hopkins in Totnes, United Kingdom, in 2005 and quickly grew into a global phenomenon. The movement's website lists over 1000 official transition groups, with the majority in Europe, Oceania, and North America [4]. The popularity of the movement has been attributed to the effective use of mass-communication technologies (books, movies, websites, and social media) to help the movement spread its message globally [5] and the emphasis on inclusiveness and pragmatism [6]. The movement advocates the transition towards low-carbon urbanism through economic localisation. The current globalised patterns of production and consumption is highly carbon intensive because of the fossil fuels consumed in the transportation of goods. The solution, then, is to prioritise the consumption of locally produced goods. Building local resilience, defined as the adaptive capacity of communities in the face of climate change, is another key objective of the transition movement. To this end, the transition movement focuses on building strong community networks, fostering social and human capital, and developing diverse skill sets.

One of the key cornerstones of the transition town movement is encouraging local food production and exchange, reflecting both the fact that food is a basic need and the movement's intellectual foundation in the permaculture movement (Hopkins was a permaculture teacher before founding the movement) [5]. One of the most established examples can be found in Totnes, UK, the first transition town. Some of the initiatives around local food production are garden sharing (an arrangement where garden owners invite fellow community members to use their garden to grow food), guerrilla gardening (growing food in public and unused spaces), seed swaps (where gardeners meet regularly to exchange seeds and knowledge), and a website to help bring together local food producers, retailers, restaurants, and consumers. In some places, the transition movement has taken localisation beyond the food economy. Most notably, some transition groups have implemented a local currency scheme to support local businesses, build closer relations between residents and local companies, and ultimately reduce carbon emissions by encouraging more local production [7]. A common criticism of conventional low-carbon approaches is that they reflect the interest of existing power structures and serve to legitimise the neoliberal paradigm. In contrast, the brand of economic localisation promoted by the transition town movement has the potential to reduce the type of embodied carbon emissions that most governments choose to ignore. The main contribution of the transition town movement to low-carbon urbanism is therefore an alternative approach that is not bounded by the conventional logic of growth and globalisation. Furthermore, the ability to involve local communities in the process means that the transition movement can be more effective than government-driven programmes, especially in soliciting changes in individual behaviours.

There are a number of limitations inherent in the movement. First, the focus on economic localisation works well in a rural, small-town setting, but adapting the objective to larger, more complex metropolitan areas has not been straightforward. For one, the scale of actions remains a contentious issue [6–8]. Any city of significant size consists of many neighbourhoods. Accordingly, should the economy be localised to the metropolitan level or the neighbourhood level? Currently, the latter approach is far more popular. London, for example, has 38 registered transition groups throughout the city are working independently from each other. However, such a segmented vision of the city contradicts the workings of the city as an integrated urban economy [6]. Second, transition groups in large cities show limited success in injecting their objectives into urban politics that are dominated by the pro-growth alliance [7]. Third, despite its attempts to be inclusive, the movement is mainly driven by white middle-class values, while disadvantaged communities are very much under-represented [7, 8]. Consequently, for it to really trigger and support broader transition from the current high-carbon trajectory of cities, the movement has yet to expand beyond a narrow demographic group—the white, middle-class, well-educated population living in relatively well-off suburbs. Fourth, the transition movement has been criticised by other environmental groups as apolitical, meaning that the movement aims to achieve social transformation through positive action rather than through antagonism [9]. While the focus on direct action is likely to contribute to the popularity of the transition movement, it runs the risk of confining the movement to irrelevance and being co-opted into the mainstream agenda [10]. Evidently, this non-confrontational approach has failed to achieve a consensus even among many local transition groups. Thus, in some cases, the approach has resulted in ideological debates, delays, loss of membership, and lack of focus [8].

Ecovillage Movement

Ecovillages are intentional communities established for the purpose of reducing environmental impacts. Ecovillages can be considered as a continuation of the green communes of the 1960s and 1970s. However, over the last two decades many new ecovillages have been established to the degree that the term ecovillage movement has been coined [11]. In 1995, the Global Ecovillage Network (GEN) was formed by 25 community representatives to promote the ecovillage concept and provide a platform for ecovillagers to share their experience. There are 347 ecovillages registered with the GEN, 147 in Europe, 48 in Oceania and Asia, and 152 in the Americas. However, the actual number of ecovillages is likely to be significantly larger because many of them are not registered with the GEN.

Ecovillages are highly diverse, which makes it difficult to generalise about them beyond the commitment to sustainable and low-carbon living. Older ecovillages are usually rural- or suburban-based and have a strong spiritual undertone, but it is increasingly common to see urban ecovillages that place more emphasis on the environmental and equality aspects of sustainability. Consequently, ecovillages are appealing to potential residents for many reasons in addition to concerns about the environment [12]. Such diversity reflects the bottom-up nature of the movement, where there is a lack of a central figure and a grand narrative, as in the case of the transition town movement.

Ecovillages set examples and standards for low-carbon lifestyle. Most ecovillages feature an impressive array of eco-technologies to improve energy efficiency and utilise renewable energy. For example, Moora-Moora, an ecovillage of approximately 30 detached homes in Melbourne, Australia, is carbon neutral and is entirely off-grid. Electricity is generated by on-site wind turbines and photovoltaic panels. Another example from the same city is WestWyck, an urban ecovillage of seven apartment units and five townhouses. Great care was taken during the construction phase to minimise embodied carbon emissions through reusing existing materials. While WestWyck is not entirely carbon neutral because of limited space, its dwellings are highly energy efficiency and are equipped with solar hot water and photovoltaic panels to significantly reduce the levels of carbon emissions.

Many ecovillages go beyond technological solutions to implement more profound ways of low-carbon and sustainable living. To many ecovillagers, a much simpler and self-sufficient lifestyle is necessary for everyone to live within the ecological limits of the planet. This means living a simpler lifestyle and sharing resources, developing a localised economy to reduce travelling to work and the transporting of products, and working less and taking time off from work to help with community projects. Not all ecovillages achieve such ideals, but many are working towards them. Moora-Moora, for example, has developed a tradition of meal-sharing to lower energy use and to promote bonding. It has also set up a car-pooling scheme to tackle the problem of high transportation emissions.

While ecovillages are primarily focused on reducing their carbon emissions and other environmental impacts through sustainable living, they are not nonchalant about the greater society. Rather, many believe that changes can best be achieved through the construction and demonstration of an operating model that can be duplicated elsewhere [13]. Take, for example, the mission statement from WestWyck:

> WestWyck has a mission to influence. Through provision of a demonstration model, it wants to support and facilitate the evolution of sustainability policies and practices that relate to the built form within an urban community.

Both Moora-Moora and WestWyck view themselves as a role model and communicate their vision to the outside world by organising open days, guided tours, conferences, and workshops. Indeed, it is increasingly common to find ecovillages actively engaging in local government, community organisations, and research institutions to spread their ideals and knowledge [14]. Non-government organisations such as Gaia Education and Living Routes are formed to facilitate the outreach process by collaborating with universities and ecovillages to offer sustainable education to students [15].

Similar to the transition town movement, the ecovillage movement offers solutions to climate change through positive action. The implicit strategy of the ecovillage movement is to start building a sustainable society here and now and to gradually increase the influence of the movement by demonstrating an alternative lifestyle to the wider society [16]. At a minimum, ecovillages contribute to low-carbon urbanism by demonstrating that it is possible, if not desirable, to live a carbon neutral lifestyle. However, there are many practical challenges the movement must overcome. An enormous amount of work is needed to establish an ecovillage, including securing financial backing, purchasing land, designing the community, and building the houses and community infrastructures. Establishing ecovillages is becoming even more difficult recently as land has become more expensive and planning control has become more complex [14, 17]. After the establishment of the ecovillage, the challenge turns to how to keep the community together. Despite the shared commitment to living sustainably, ecovillages are made up of individuals of heterogeneous personalities, backgrounds, priorities, expectations, values, and interests. Moreover, conflicts may arise between residents who contribute more and those who contribute less to the community because of other responsibilities (jobs, education, children, and so on) [18]. As a result, the communal aspect of ecovillages, rather than being the glue, often has become the source of internal conflict that, if handled inappropriately, may tear the ecovillage apart.

Community Renewable Energy

Community renewable energy (CRE) has grown in popularity over the past decade [19]. Compared to the traditional, centralised, carbon intensive form of power generation, locally based distributed renewable power has significant potential to decarbonise the power sector, improve the resilience of the system against extreme disturbances, and provide a tangible experience of energy production that is beneficial for energy education and the promotion of environmental citizenship [20]. CRE also gives the community greater economic security and generates a financial return to the community. CRE, being focused on a particular kind of eco-technology, is more specific than the transition movement and the ecovillage movement. Nevertheless, there is significant diversity in terms of renewable energy technologies, ownership models, and organisational structure (e.g., cooperatives, community charities, or development trusts). In Oxfordshire, United Kingdom, for instance, Westmill Co-operative was established in 2004 by a group of local residents to create and operate a community-owned wind farm. Westmill successfully attracted 2374 members and a share capital of £4.6m. With additional bank loan, the co-op purchased and installed five 1.3 MW wind turbines. Encouraged by the success, Westmill Solar Co-operative was recently formed to develop a 5 MW community-owned solar park with over 20,000 polycrystalline photovoltaic panels, enough to generate 4.8 GWh of electricity a year, or prevent 2000 tCO_2 emissions annually. In another example, Solarize Portland is a community campaign for collective purchasing of residential photovoltaic systems. The campaign began in 2009 when two local residents started a local bulk-purchasing scheme to reduce the cost and complexity of residential solar power [21]. Compared to the failure of previous municipality-led schemes to promote solar power installations, Solarize Portland was very successful and in its first year helped over 120 households to purchase and install solar power systems.

The effective use of community resources is one of the reasons for the success of CRE projects. CRE relies on the mobilisation of local social ties to generate demand for technological change, and as more people join in, creates a 'virtuous social-technical loop' that helps to push the project forward [21]. Furthermore, CRE projects are generally more acceptable to the community than large-scale private investments because they are generally more locally appropriate, involve more local people in the process, and bring more benefits to the locality through local ownership [22, 23]. Therefore, whereas the development of large-scale commercial wind farms has proceeded slowly in many countries because of strong local opposition [24, 25], community wind farms are mushrooming in many countries, including Australia, Japan, Europe, and North America [20, 26–28]. CRE projects are also associated with a number of challenges, including setting up local distribution networks, the long-term capacity of community organisations to maintain and operate systems efficiently, overcoming barriers to market entry and network connection, securing loans, and obtaining green energy certificates [20, 29].

Conclusion

The influence of grassroots environmentalism on low-carbon urbanism is growing but remained highly fragmented and localised, as exemplified by the transition town movement, the ecovillage movement, and community renewable energy initiatives. The transition movement focuses on building low-carbon and resilient communities in a post-oil society. The ecovillage movement demonstrates the viability of a sustainable and low-carbon lifestyle. Community-based projects can be effective in promoting the deployment of renewable energy at a local scale. Like many other grassroots initiatives, community low-carbon action has a strong commitment to the principles of community collectivism, such as community visioning, inclusion and diversity, and building partnership [1]. Grassroots low-carbon initiatives are therefore valued not only for their environmental benefits but also for a range of social benefits, such as community empowerment and the fostering of social capital.

There has, to date, been no measurement of the impacts of grassroots low-carbon initiatives in terms of reducing carbon emissions and energy consumption, but debates have been waged on their political impacts. While low-carbon community action is regarded by some as a marginal and quixotic social phenomenon, it nevertheless advances low-carbon urbanism by introducing progressive logics for addressing climate change and energy security. In particular, the focus on community building, resilience, localisation, and dematerialisation constitute an obvious point of departure from the mainstream, state-sponsored action that can be considered as a continuation of an existing ideology of environmental liberalism. Rather than business-as-usual, advocates of low-carbon community action envision a decarbonised, localised, and non-consumerist society where trade will be minimised as much as possible and people will practice a sustainable, community-focused, simple way of living: a fundamental transformation of everyday practices.

Despite some successes, there are many challenges with regard to adapting the creative energies of grassroots initiatives for wider mainstream settings [30]. Grassroots initiatives by definition rely on people with limited power, resources, and ability to influence others [31]. The impact of low-carbon community action therefore depends on its ability to address these issues through capacity building. Capacity building can be top-down, such as through supportive policies and legal frameworks, or bottom-up, such as by forming partnerships and fundraising. At the same time, because grassroots initiatives are often formed in opposition to unsustainable regimes, community activists may worry about the incorporation into mainstream contexts [30]. The ways through which the tension between growth and mainstreaming are negotiated are therefore also a critical factor for the long-term development of grassroots community low-carbon action.

References

1. Hopkins, R.: The Transition Companion: Making Your Community More Resilient in Uncertain Times. Chelsea Green Publishing, White River Junction (2011)
2. Bulkeley, H.: Cities and Climate Change. Routledge, Oxon (2013)
3. Lo, K.: Urban carbon governance and the transition towards low-carbon urbanism: review of a global phenomenon. Carbon Manag. **5**, 269–283 (2014)
4. Transition Network: www.transitionnetwork.org
5. Bailey, I., Hopkins, R., Wilson, G.: Some things old, some things new: the spatial representations and politics of change of the peak oil relocalisation movement. Geoforum. **41**(4), 595–605 (2010)
6. Taylor, P.J.: Transition towns and world cities: towards green networks of cities. Local Environ. **17**(4), 495–508 (2012)
7. North, P., Longhurst, N.: Grassroots localisation? The scalar potential of and limits of the 'transition' approach to climate change and resource constraint. Urban Stud. **50**(7), 1423–1438 (2013)
8. Smith, A.: The transition town network: a review of current evolutions and renaissance. Soc. Mov. Stud. **10**(1), 99–105 (2011)
9. Bay, U.: Transition town initiatives promoting transformational community change in tackling peak oil and climate change challenges. Aust. Soc. Work. **66**(2), 171–186 (2013)
10. Connors, P., McDonald, P.: Transitioning communities: community, participation and the Transition Town movement. Commun. Dev. J. **46**(4), 558–572 (2011)
11. Jackson, R.: The ecovillage movement. Permaculture Mag. 1–11 (2004)
12. Kasper, D.V.S.: Redefining community in the ecovillage. Res. Hum. Ecol. **15**, 12–24 (2008)
13. Kirby, A.: Redefining social and environmental relations at the ecovillage at Ithaca: a case study. J. Environ. Psychol. **23**(3), 323–332 (2003)
14. Dawson, J.: From islands to networks: The history and future of the ecovillage movement. In: Lockyer, J., Veteto, J.R. (eds.) Environmental Anthropology Engaging Ecotopia: Bioregionalism, Permaculture, and Ecovillages. Berghahn Books, Oxford (2013)

15. Greenberg, D.: Academia's hidden curriculum and ecovillages as campuses for sustainability education. In: Lockyer, J., Veteto, J.R. (eds.) Environmental Anthropology Engaging Ecotopia: Bioregionalism, Permaculture, and Ecovillages. Berghahn Books, Oxford (2013)
16. Trainer, T.: Where are we, where do we want to be, how do we get there? Democr. Nat. **6**(2), 267–286 (2000)
17. Lockyer, J.: Intentional community carbon reduction and climate change action: from ecovillages to transition towns. In: Peters, M., Fudge, S., Jackson, T. (eds.) Low Carbon Communities: Imaginative Approaches to Combating Climate Change Locally, p. 197. Edward Elgar Publishing, Cheltenham, UK (2010)
18. Baker, T.: Ecovillages and capitalism: creating sustainable communities within an unsustainable context. In: Lockyer, J., Veteto, J.R. (eds.) Environmental Anthropology Engaging Ecotopia: Bioregionalism, Permaculture, and Ecovillages. Berghahn Books, Oxford (2013)
19. Toke, D.: Community wind power in Europe and in the UK. Wind Eng. **29**(3), 301–308 (2005)
20. Adams, C., Bell, S.: Local energy generation projects: assessing equity and risks. Local Environ. **20**(12), 1473–1488 (2015)
21. Aylett, A.: Networked urban climate governance: neighborhood-scale residential solar energy systems and the example of Solarize Portland. Environ. Plann. C Gov. Policy. **31**, 858–875 (2013)
22. Warren, C.R., McFadyen, M.: Does community ownership affect public attitudes to wind energy? A case study from south-west Scotland. Land Use Policy. **27**(2), 204–213 (2010)
23. Walker, G.: The role for 'community' in carbon governance. Wiley Interdiscip. Rev. Clim. Change. **2**(5), 777–782 (2011)
24. Pepermans, Y., Loots, I.: Wind farm struggles in Flanders fields: A sociological perspective. Energy Policy. **59**, 321–328 (2013)
25. Hall, N., Ashworth, P., Devine-Wright, P.: Societal acceptance of wind farms: analysis of four common themes across Australian case studies. Energy Policy. **58**, 200–208 (2013)
26. Hicks, J., Ison, N.: Community-owned renewable energy (CRE): opportunities for rural Australia. Rural Soc. **20**(3), 244–255 (2011)
27. Walker, G., et al.: Harnessing community energies: explaining and evaluating community-based localism in renewable energy policy in the UK. Glob. Environ. Polit. **7**(2), 64–82 (2007)
28. Maruyama, Y., Nishikido, M., Iida, T.: The rise of community wind power in Japan: enhanced acceptance through social innovation. Energy Policy. **35**(5), 2761–2769 (2007)
29. Walker, G.: What are the barriers and incentives for community-owned means of energy production and use? Energy Policy. **36**(12), 4401–4405 (2008)
30. Seyfang, G.: Community action for sustainable housing: building a low-carbon future. Energy Policy. **38**(12), 7624–7633 (2010)
31. Middlemiss, L., Parrish, B.D.: Building capacity for low-carbon communities: the role of grassroots initiatives. Energy Policy. **38**(12), 7559–7566 (2010)

Kevin Lo is an Assistant Professor at the Department of Geography in Hong Kong Baptist University. He is a human geographer with a research focus on the development of effective policy interventions and governance mechanisms to promote renewable energy and energy efficiency. He has published in many leading journals, including Renewable and Sustainable Energy Reviews, Energy Policy, Energy for Sustainable Development, Energies, Environmental Science & Policy, Habitat International, and Cities.

Emerging Low-Carbon Urban Mega-Projects

Federico Caprotti

Abstract This chapter focuses on the recent trend in some geographical locations (particularly China and the United Arab Emirates) towards building large-scale low-carbon city projects. These low-carbon cities are increasingly being described as mega-projects due to their scale and involvement of large-scale experimental approaches to ways of organizing the low-carbon city. The chapter discusses some of the main trends towards the development of low-carbon eco-city projects since 2000, and then introduces the two main low-carbon city mega-projects currently being developed.

Keywords Eco-cities • Urban development • Urban geography • Eco-urbanism • Green economy

Key Terms

Eco-cities. Cities (usually new urban projects) which aim to be environmentally, economically and socially sustainable

Eco-urbanism. A range of approaches to the city which try to render it more environmentally, economically and socially sustainable

1 Introduction

The development of low-carbon urbanism in the twenty-first century has become a central focus of urban planning and policy at the international and national scales. The key crises of the twenty-first century—from the climate crisis, to rapid urbanization, urban health, economic sustainability, and even security—place the city at the centre of debate. Eco-cities have become especially popular ways of thinking about changing the urban environment, or building completely new cities and urban

F. Caprotti (✉)
Department of Geography, University of Exeter, Amory Building, Rennes Drive,
Exeter EX4 4RJ, Devon, UK
e-mail: f.caprotti@exeter.ac.uk

S. Dhakal, M. Ruth (eds.), *Creating Low Carbon Cities*,
DOI 10.1007/978-3-319-49730-3_6

districts, so as to orient cities in such a way that they become better able to face the risks identified above. Specifically, new low-carbon urban developments are aimed at achieving two distinct aims: ensuring the development of more environmentally sustainable, lower-impact urban areas, and stimulating the growth of a low-carbon economic basis around low-carbon urban projects. By far the most visible and capital-intensive individual low-carbon urban developments since 2000 have been eco-city projects.

In part, there is nothing new about eco-cities when viewed in the context of sustainable urban development. The attempt to integrate concerns over sustainability and to build more environmentally harmonious urban environments has been central to urban planning in a range of national contexts since at least the end of the nineteenth century, with the popularity and planning influence gained by Ebenezer Howard's Garden City concept. In turn, the twentieth century saw a range of experimental New Towns, some of which placed nature and the environment at the centre of their urban plans. This was the case in localities as diverse as Canberra, Australia; Curitiba, Brazil; the 'green' towns of the New Deal in the USA in the 1930s; and many of the New Towns built in Italy under Mussolini. Thus, notions of the reconciliation of societal, economic and environmental 'goods' in newly planned cities has been a constant ingredient in master plans and urban blueprints for well over a century.

While environmentally sustainable urban development has been a relatively common feature in the urban planning landscape since the end of the 1800s, there are distinctive features and characteristics to eco-city developments in the 2000s. One of the distinguishing features of current eco-city and other low-carbon urban developments is their experimental nature [1]. This means that eco-city projects constructed since 2000 largely display characteristics of experimentation—with infrastructure, urban form, transport, material flows, consumption patterns, governance and the like (see [2] for categories of urban experimentation)—which conceptualize the city as an arena for experimentation, innovation, and trial and error.

One of the key distinguishing features of eco-city developments in the 2000s is the involvement of both state and corporate actors in the envisioning, planning and delivery of eco-city projects. This is particularly evident in a range of projects, especially outside Europe and North America, where corporations are central to every stage of project planning and implementation. This includes the development of master plans by global engineering and urban planning consultancies; the design of individual buildings by architectural firms; the evaluation and monitoring of projects by environmental and other auditing corporations; and the development and marketing of new cities' residential areas by real estate development companies. This has meant that while state actors (from ministers, to policymakers, to urban planners, to sovereign wealth funds) are key drivers behind new urban projects, new eco-cities are no longer conceptualized without significant private sector involvement. This leads to questions around the politics of new eco-cities, which will be discussed in more detail at the end of the chapter.

When new urban areas are being designed, built and developed through state–corporate partnerships, what is easily left out of new city plans is the role of 'lay' knowledge and preference. In particular, it appears that new eco-cities carry with them significant socio-political risk because if new cities are designed by high-level state and corporate elites, then the city as the result of a *political process* ceases to exist, potentially leading to the formation of 'post-political' urban environments [3].

Furthermore, eco-city developments since 2000 are characterized by an increasing number of new projects being started in the Global South, and especially in rapidly emerging economies such as India and China. This points to the development of experimental urbanism in societal, economic and political contexts which are in many cases far removed from previous rounds of new-built, experimental urban developments, which tended to be predominantly focused on Europe and North America. The fact that most of the large-scale, mega-project urban development currently underway is *not* taking place in Europe and North America is not just a reflection of the higher rates of urbanization in emerging economies. It signifies a more radical and comprehensive shift towards these economies when cities as a whole are—for better or worse—perceived as up for grabs and open to radical change and development.

Moreover, the political and cultural arena which has in part enabled the planning and development of eco-cities has done so whilst depicting and constructing eco-cities largely as technological, ecologically modernizing responses to constructed notions of *crisis* [4]. What 'crisis' means in each national and urban context is different, but one of the most common discursive rationales presented for the need to invest in eco-city development is the notion that new urban areas need to be built as a result of ongoing fears about environmental crisis, environmental change, and rising sea levels. While such concerns have been mainstream since at least the 1960s, rising levels of global urbanization have placed cities centre-stage as the interface where society and the environment meet. In addition, global UN conferences and international agreements on the environment have focused attention on the role of cities in a context of changing climate and increasing environmental risks to cities. For example, in a survey of urban climate change experimental projects, it was found that 79 % of such projects were initiated after the 2005 ratification of the Kyoto Protocol:

> This is not necessarily an indication that international agreements have direct impact in fostering climate change experimentation, but rather, that international climate change governance efforts correspond with an increasing interest on climate change in the collective imaginations of urban actors. Climate change has gained more visibility in the city at the same time as the agreements took place [5].

Eco-cities have also been constructed as adequate responses to other types of crisis, notably that of 'Peak Oil': many of these new urban projects are presented as new 'ways of doing' and as novel approaches to the economy, not based as centrally on oil as other urban and heavily industrial areas. This does not of course mean a rejection of the market, for in ecologically modernizing eco-city projects the market is a central actor, and environmental technologies and renewable energies are the new commodities and industrial products of a decarbonizing economy.

In short, the development of low carbon eco-cities since 2000 is a culturally, economically, ideologically and politically specific project that is being articulated across a range of geographies and contexts: it is in this sense that Joss [6] has argued that eco-cities have now become mainstream. The next section considers the landscape of eco-city projects in more detail.

2 A Geography of Eco-Cities

Providing an overview of eco-city developments worldwide is difficult. This is in part due to the wide variety of eco-cities that can be found globally, and because of the rapid emergence of eco-city projects since 2000. In terms of the range of eco-city developments to be found across the globe, recent surveys by the International Eco-Cities Initiative headquartered at the University of Westminster, London, identified 174 eco-city projects at various stages of development, from the master-planning stage to completed projects [7]. These are geographically distributed across the world, although Asia and Australasia (with 69 projects) and Europe (70 projects) dominate in numerical terms, compared with Africa and the Middle East (10 projects) and North, Central and South America (25 projects).

In turn, the overall landscape of eco-city development worldwide needs to be set in a context which considers the typology of projects commonly referred to as 'eco-cities'. This is because there does not exist a single, unitary definition of what an eco-city actually is, or should be. Rather, the eco-city concept has been used to label and describe a wide range of projects and urban interventions aimed at making urban areas more ecologically sensitive. Therefore, the wide geography of eco-city developments outlined above necessarily includes projects which are very different in scope and scale: from mega-projects, to cities focusing on a specific technology, to neighbourhood redevelopments and eco-districts. Building on the analysis by Joss, Tomozeiu and Cowley mentioned above, the following categories into which eco-city projects fall can be used: (a) retro-fitting of existing urban areas, at a variety of scales (from neighbourhoods, to brownfield sites, to a range of individual buildings linked in an urban whole); (b) expansions of already existing cities; (c) newly built eco-cities which are not expansions of existing cities but completely new urban areas in and of themselves. Thus, eco-cities range from small neighbourhood developments (such as BedZED in London), to whole new cities, such as Treasure Island, an eco-island in California [8].

The global geography of low-carbon eco-city projects is one characterized by diversity in size, scope and stage of development, as well as by differences in the scale and scope of such developments. In terms of scope, there is further heterogeneity in the range of low-carbon urban projects worldwide. Some low-carbon city projects are based around the experimental use of a single technology or technological sector. For example, the city of Rhizao, in Shandong province in southern China, has become a case study for the implementation of technological policies aimed at the use of solar heating and energy. The city is seen as an example of the positive results of the combined collaborative agency of government and private sector actors at a variety of levels. At the national level, government policies promoting the solar manufacturing sector in China had an effect on local solar firms in and around Rizhao, providing incentives for research, development and manufacturing. In turn, Rizhao's municipal government was active in linking with the national policy sphere and opening the city to the adoption of newly manufactured solar technologies. Some of these technologies were not high-tech or cutting edge, such as solar heaters. However, the positive adaptive environment engendered by state, private sector and municipal actors led to a situation whereby 99 % of buildings in the city centre used solar water heaters by the end of the 2000s [17], and most municipal energy-using infrastructure (for example, traffic lights and streetlights) was powered through photovoltaic (PV) panels supplied by local solar manufacturing corporations.

While some low-carbon urban projects are focused on a single set of technologies, others are more ambitious and wide-ranging in their conceptualization of the role of the city vis-à-vis the carbon economy. Some low-carbon eco-city projects are increasingly deploying 'Smart City' concepts, for example, in order to conceptualize urban areas in terms of complex systems where efficiencies in energy and resource use can be achieved by harnessing the city's networked potential. Such approaches mean, for example, the construction or retrofitting of cities so that they become interconnected, digitally enabled networks of information and command functions. This is the case, for example, with cities using Smart Grid networked technologies to try and ensure more efficient energy distribution and use in the urban area, and less waste and over- and under-supply of energy.

An example of a low-carbon project using a Smart City approach is Songdo City. Songdo is a new project in South Korea near Incheon: it features a Central Business District (CBD) planned so as to integrate networked infrastructure into the material space of the city. Termed an 'innovative city for a utopian future' [9], the city features informational hardware and software aimed at improving energy and resource use in the urban area. This includes a pneumatic garbage collection system, which suctions waste directly from apartments into a central waste processing plant; sensors to monitor traffic flows, energy use and temperatures so as to enable management systems to react to the city's changing environment; and a network of electric car charging points throughout the CBD so as to encourage greener forms of transport. The aim of these smart city technologies is to build a synergy between the 'digital city' and the 'eco-city', using networked technologies as ways of making the city as environmentally friendly and sustainable as it can be.

The two approaches to eco-cities above (based on a single set of technologies or on organic approaches to integrating new socio-technical systems into the urban system) are technology-focused and site-focused. What must be mentioned when considering the *geography* of eco-cities, however, is their distribution vis-à-vis the 'problems' (from environmental degradation to Peak Oil and rural-urban migration) that these urban environments are meant to 'solve'. Many of the impacts of climate change will inordinately impact the world's poorest populations.

Many of the most concentrated impacts will be in urban areas, many of which are organic urban agglomerations which are highly exposed to the climatic, environmental and associated socio-economic and health hazards of climate change. For example, the Climate Change Vulnerability Index (CCVI), part of the recently published fifth edition of the *Climate Change and Environmental Risk Atlas* [10], highlights seven cities at the highest risk of climate change-related risks: Dhaka, Bangladesh; Manila, Philippines; Bangkok, Thailand; Yangon, Myanmar; Jakarta, Indonesia; Ho Chi Minh City, Vietnam; and Kolkata, India. Recent events such as major flooding in Bangkok in 2011, and Typhoon Haiyan, which devastated the Philippines in November 2013, highlight the climate and meteorological vulnerability of urban areas in less developed contexts.

However, when considering a geography of eco-cities, it is interesting (and alarming) to note that very few eco-city projects are taking place in the poorest parts of the world: those parts where urban dwellers will suffer the most from climate change. For example, as the International Eco-Cities Initiative global survey highlighted [7], only 10 out of 164 eco-city projects worldwide are in Africa and the Middle East. Of these, only three projects are in less developed countries south of the Sahara Desert (the projects are in Kenya, Nigeria and Uganda). Of the 69 projects in Asia and Australasia, most are in developed countries (such as South Korea or Japan) or newly emerging economies (such as China). Very few are located in countries which are less developed. In the case of Indonesia and the Philippines—two countries at great risk of climate-induced impacts—there is a single eco-city project in each of these countries, although in the case of the former, the plan is for a network of eco-cities. This in turn leads to pressing questions around the future of cities in the least economically developed areas of the world, and of what can be done at an international and national level to make sure that some sort of adaptive eco-urban planning takes places in cities in these locations.

3 Eco-City Mega-Projects

Apart from low carbon eco-city projects based around a circumscribed set of technologies or around the low carbon economy, there are specific, large-scale attempts to construct large, newly built eco-cities which focus both on the sustainability of the urban environment, and on the establishment of a low-carbon, green economy basis to socio-economic activity both within and outside the eco-city proper. Eco-city mega-projects are defined here as completely newly built eco-cities based on blueprints and master plans which attempt to fashion a new vision of the city in relation not only to a low-carbon economic basis, but also in relation to building the city as a resilient socio-economic system in light of the challenges of a changing climate and environment. Eco-city mega-projects are therefore defined not only as low-carbon, but as embedding resilience within the master-planning of the city. In addition, eco-city mega-projects are defined here as new eco-city developments which are planned for populations of 20,000 or more.

This helps distinguish mega-projects from the retro-fitting of urban areas, or from smaller-scale newly built projects such as BedZED in south London, or the Vauban neighbourhood in Freiburg, Germany. It is important to acknowledge the fact that eco-city mega-projects are relatively few compared to the large number of smaller eco-city projects currently planned or under construction. Nonetheless, they are worthy of study due to their scope, scale and importance in terms of government and private sector involvement, and in terms of the flows of capital, technological know-how and expertise which are focused on the development of these mega-projects.

Two eco-city mega-projects currently stand out globally in terms of scale, scope, vision and stage of development and construction. These are Masdar Eco-City, in Abu Dhabi, United Arab Emirates (UAE); and the Sino-Singapore Tianjin Eco-City, near Tianjin and Binhai, China. Both of these projects are new-build and large-scale, and both are at advanced levels of construction at the time of writing.

Masdar Eco-City

In the case of Masdar, the eco-city was founded in April 2006 with a clear transitional focus, aimed at providing Abu Dhabi with a new city with experimental approaches to a low-carbon economy. This is intended to help transition the Abu Dhabi economy away from a reliance on oil, to a more diversified economic basis including environmental and renewable energy technologies. This is a transitional strategy not only in urban and technological terms, but also in terms of enabling the current, high-value human capital in Abu Dhabi to transition into constructing a new high-tech research, development and commercialization hub which leverages the intellectual and technical know-how associated with the oil industry and applies it to the development of a low-carbon economy in Abu Dhabi. Masdar Eco-City is central to this aim, most visibly through the siting of a new educational institution on the eco-city site. The Masdar Institute of Science and Technology (MIST) is the result of a partnership with the Massachusetts Institute of Technology (MIT) and is the first educational institution which offers programmes of study wholly focused on renewable energy and clean technologies. Opened in 2009, MIST is part and parcel of the eco-city mega-project's aim of stimulating a high-value, low-carbon economy in the area in and around Masdar.

Although Masdar was founded in 2006, the ground-breaking ceremony actually took place in 2008. The project represents a significant capital investment on behalf of the Mudabala Development Company, an investment corporation owned by the Abu Dhabi government. The total project cost is of over US$20 billion, although the final cost at project completion is unknown, and the target date for completion was initially 2015, then 2025, and is now rather vague.

The investment of large amounts of capital into the Masdar project by the Abu Dhabi government can clearly be seen as an example of a targeted investment not only in a low-carbon urban area, but in a (future) low-carbon economy. Indeed, the master plan calls for Masdar to function as the centre of an economic-industrial hub represented by a Special Economic Zone focusing exclusively on firms within the green economy. The Special Economic Zone attempts to attract corporations and investors in this economic vision through the abolition of taxes and import tariffs for clean technology firms and associated corporations. There is also no restriction on capital movement into and out of the Special Economic Zone, making it easy, efficient and quick to invest in Masdar and its low-carbon economic zone.

In urban planning terms, the Masdar site comprises 6 km^2 of land outside the city of Abu Dhabi proper, and is scheduled to house up to 50,000 residents, although by 2013 only 200 individuals had been able to move to the eco-city: all of them were students at MIST. International architecture and urban design firm Foster + Partners was responsible for designing the urban plan for the eco-city, which features a wall around its perimeter to keep out the desert wind: this feature has also prompted the criticism that Masdar may turn into a large-scale example of a gated community, a 'premium ecological enclave' [11]. In addition, low-carbon innovations in Masdar include high-tech solutions such as the use of solar energy to power the city, wind towers to provide street-level ventilation, and personal transport pods running along tracks laid throughout the city (Masdar is meant to be carless, and linked to Abu Dhabi via a rapid transport link, although an open question is to what extent Abu Dhabi's citizens will want to use public transport to travel to and from the eco-city). Furthermore, the plan for Masdar vis-à-vis its carbon use has been modified: the initial master plan called for the eco-city mega-project to be carbon neutral, but in the wake of the 2008 financial crisis the eco-city's aim was changed to being 'low carbon'. Another open question is the extent to which Masdar can not only be low-carbon, but also socially sustainable and just [12]. Indeed, if the eco-city is developed as a residential centre exclusively for those working in high-tech industries, then Masdar risks becoming a 'premium ecological enclave' not just aesthetically and in architectural and urban design terms, but in socio-economic terms as well.

Sino-Singapore Tianjin Eco-City

Sino-Singapore Tianjin Eco-City is an eco-city mega-project currently being constructed in China, between the cities of Tianjin and Binhai. The city, generally known as Tianjin Eco-City or by its acronym (SSTEC), is the largest eco-city mega-project of its kind. The eco-city is being built on a former wetland site, which is problematic in itself, and the project calls for an eco-city eventually housing 350,000 people. By the end of 2013, nearly 10,000 residents had moved in to the city's Start-Up-Area (SUA), making the city the most significant eco-city development in the world in terms of having moved the farthest from the drawing board to reality.

SSTEC is structured as a joint venture between the Chinese and Singaporean governments, and is backed by the leadership of both countries. Both China and Singapore own 50 % of the consortium which is charged with developing and building the eco-city, and therefore the project itself comes under the almost direct ownership of the countries' governments. While the plans and funding for the project are government-led, SSTEC also involved significant private sector involvement, especially by real estate development corporations (such as Singapore's Keppel Corporation, Taiwan's Farglory or Malaysia's Sunway).

The eco-city's ground-breaking ceremony took place in September 2008, after an eco-city selection process which evaluated several sites throughout China and which took around a year. The selection of Tianjin municipality as a site for this flagship eco-city was significant, in that Tianjin is one of China's major ports and industrial and cargo gateways to the world. It is also located within an hour's train journey to central Beijing, and is thus close to the centre of China's political power, symbolizing the importance of the eco-city project for the Chinese government. Indeed, the political importance of the SSTEC project can be seen by the fact that the ground-breaking ceremony was led by Wen Jiabao, then the Chinese premier, and Goh Chok Tong, the Senior Minister of Singapore.

The master plan for the eco-city calls for a city surrounded, as in the case of Masdar, by a Special Economic Zone which favours environmental technology and services firms, including the manufacturing of environmental technology products for export through Tianjin cargo port, one of the largest in the world. The use of an experimental Special Economic Zone is not new in the Chinese context: since the development of Shenzhen as an economic hub in the 1970s, Special Economic Zones have been the instrument of choice for stimulating new forms of economic and industrial growth in China. Nonetheless, the economic zone around Tianjin is interesting in that it aims to promote a transition towards a high-tech, low-carbon economy in the midst of a wider regional context which sees Tianjin as the central location in a region characterized by heavy industries and manufacturing.

The urban planning and design for the low-carbon eco-city called for an initial SUA to be developed. This zone, partially completed by 2013, called for 26,500 apartments and 85,000 residents by 2013, although fewer than 10,000 individuals had moved in by 2013. The SUA is a significant parcel of land in the eco-city, since it comprises an area of nearly 8 km^2. By 2033, the eco-city is expected to house 350,000 residents in around 110,000 apartments. The eco-city's master plan, developed by the China Academy of Urban Planning and Design, the Tianjin Urban Planning and Design Institute, and a Singaporean planning team under the aegis of Singapore's Urban Redevelopment Authority, calls for the integration of low-carbon innovations within the eco-city mega-project. These include the partial powering of the city by solar and wind energy; the use of a pneumatic garbage collection system; district heating; and the inclusion of advanced water filtration and recycling systems in the city's infrastructural organization.

While SSTEC is clearly an innovative mega-project, several questions remain as to whether it goes far enough in its ambitions. For example, questions remain as to the level of carbon use by the eco-city upon project completion, especially since few of the eco-city's buildings attain the highest international standards of energy conservation and efficiency. Furthermore, the eco-city clearly features large, multi-lane roads separating residential blocks, leading to questions about whether the city can adequately and successfully implement policies aimed at promoting the use of

4. Caprotti, F.: Eco-urbanism and the eco-city, or, denying the right to the city? Antipode. **46**(5), 1285–1303 (2014)
5. Castán Broto, V., Bulkeley, H.: A survey of urban climate change experiments in 100 cities. Glob. Environ. Change. **23**(1), 96 (2013)
6. Joss, S.: Eco-cities: the mainstreaming of urban sustainability: key characteristics and driving factors. Int. J. Sust. Dev. Plann. **6**(3), 268–285 (2011)
7. Joss, S., Tomozeiu, D., Cowley, R.: Eco-Cities: A Global Survey 2011. University of Westminster International Eco-Cities Initiative, London (2011)
8. Joss, S.: Eco-city governance: a case study of Treasure Island and Sonoma Mountain Village. J. Environ. Policy Plann. **13**(4), 331–348 (2011)
9. Shwayri, S.T.: A model Korean ubiquitous eco-city? The politics of making Songdo. J. Urban Technol. **20**(1), 39–55 (2013)
10. Maplecroft: Climate Change and Environmental Risk Atlas, 5th edn. Maplecroft, Bath (2013)
11. Hodson, M., Marvin, S.: Urbanism in the anthropocene: ecological urbanism or premium ecological enclaves? City Anal. Urban Trends Cult. Theory Policy Action. **14**(3), 299–313 (2010)
12. Cugurullo, F.: How to build a sandcastle: an analysis of the genesis and development of Masdar City. J. Urban Technol. **20**(1), 23–37 (2013)
13. Chien, S.S.: Chinese eco-cities: a perspective of land-speculation-oriented local entrepreneurialism. China Info. **27**(2), 173–196 (2013)
14. Wong, T.C.: Eco-cities in China: pearls in the sea of degrdaing urban environments. In: Wong, T.-C., Yuen, B. (eds.) Eco-City Planning: Policies, Practice and Design, pp. 131–150. Springer, New York (2011)
15. Rapoport, E.: Utopian visions and real estate dreams: the eco-city past, present and future. Geogr. Compass. **8**(2), 137–149 (2014)
16. Joss, S., Cowley, R., Tomozeiu, D.: Towards the 'ubiquitous eco-city': an analysis of the internationalisation of eco-city policy and practice. Urban Res. Pract. **6**(1), 54–74 (2013)

References: Websites

17. Bai, X.: China's solar powered city. Renewable Energy World, 22 May 2007. http://www.renewableenergyworld.com/rea/news/article/2007/05/chinas-solar-powered-city-48605 (2007). Accessed 1 Nov 2013

Federico Caprotti is an associate professor in human geography at the University of Exeter. As an urban geographer, Dr. Caprotti's main research interests are in nature and the city, and especially in the way in which environmental issues are constructed and deployed politically and ideologically in a range of contexts. He gained his doctoral degree at Oxford University's School of Geography and the Environment.

Energy Consumption and Emissions Assessment in Cities: An Overview

Lucia Alexandra Popartan and Francesc Morata

> How cities develop will determine our collective ability to address climate change [...]. The time has come to bring the experience and the capacity of cities to deal with the development and climate change challenge to the front of the climate debate. We need to empower cities to do the right things on climate change and we need to learn from their experience.
>
> *Angel Gurría, OECD Secretary General*

Abstract Cities are major consumers of energy and, at the same time, energy consumption is the largest contributor to carbon dioxide (CO_2) emissions. Nowadays, more than half of the world population is living in cities and this percentage is expected to continue increasing rapidly due to growing urbanization in emerging economies. This tendency places cities at the centre of the sustainable energy challenge and therefore the policies aimed at ensuring reliable energy supply and sustainable energy generation at the local level acquire increased relevance. However, there are huge disparities among cities, and measuring the degree to which cities contribute to climate change is not an easy task. This chapter provides a snapshot of current trends in energy consumption and emissions at the urban level, followed by an overview of the complexity of providing reliable and fit-for-use inventories. It also discusses the degree to which cities can be viewed as a solution instead of a problem, given the wide range of actions that can be implemented at the urban scale in order to minimize energy use and adapt to climate change.

L.A. Popartan (✉)
LEQUIA, Institute for the Environment, Universitat de Girona, Faculty of Science, Campus Montilivi, 17071 Girona, Catalunya, Spain
e-mail: alexandra.popartan@gmail.com

F. Morata
EUgov, Research Group on Multilevel Governance in the EU, Universitat Autonoma de Barcelona, Bellaterra, Spain

S. Dhakal, M. Ruth (eds.), *Creating Low Carbon Cities*,
DOI 10.1007/978-3-319-49730-3_7

Keywords Urbanization • GHG emissions inventories • Ecologic footprint • Energy consumption • Climate change adaptation

Acronyms

GHG	Greenhouse gases
CO_2	Carbon dioxide
ICLEI	Local Governments for Sustainability
UN	United Nations
UNEP	United Nations Environment Programme
UNECE	United Nations Economic Commission for Europe
UNFCCC	United Nations Convention on Climate Change

1 Introduction

Nowadays, cities are the dominant form of human settlement. They are voracious consumers of energy, which in turn generates large amounts of CO_2 emissions, placing cities at the forefront of the climate change debate. Two key elements determine the central role of cities in climate action. In the first place, cities cover only 3 % of the global surface but consume between 60 and 80 % of the total commercial energy [1]. Since global energy demand is dominated by fossil fuels, 70 % of the world greenhouse gas (GHG) emissions have been attributed to cities [2]. Secondly, urban centres are exposed to considerable destructive effects of climate change such as extreme weather events and rising sea levels.

This chapter offers an overview of current energy consumption and GHG emissions at the urban level, emphasizing the complexity of measuring the environmental footprint of cities. Moreover, it also discusses the role cities can play in alleviating climate change effects: thanks to the concentration of population, of economic and cultural activity, cities hold an enormous potential to generate knowledge and innovation to the service of climate change adaptation efforts. Moreover, local leaders are generally best suited to design strategies to address their infrastructure needs, land use, geography, and economic profiles that can have positive effect on the reduction of energy demand and GHG emissions.

Energy Consumption, GHG Emissions and Urbanization

Urbanization and industrialization were the two main traits of economic development in the early twentieth century in Europe, North America and Japan. While urbanization has gradually slowed down in these regions, the process is now

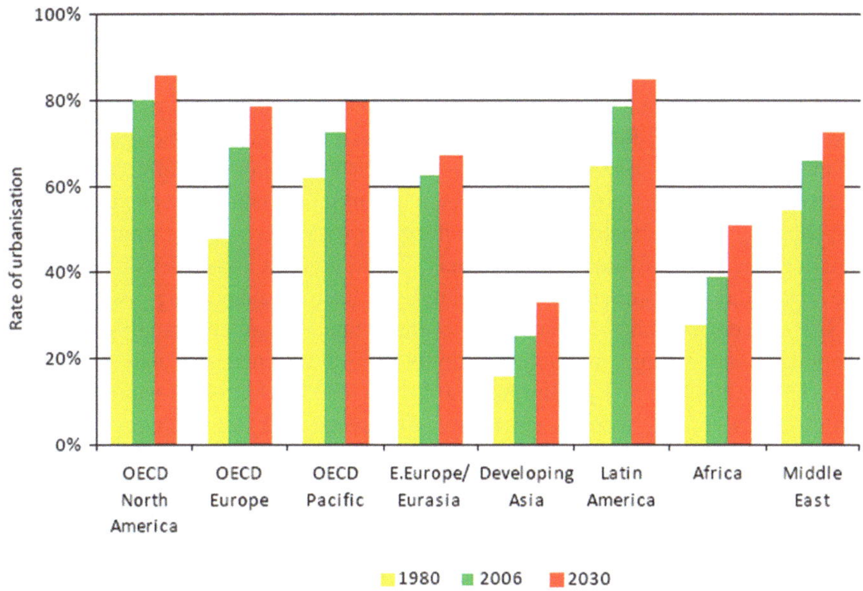

Fig. 1 Urbanization trends. Source: UN Economic Commission for Europe, [5]

witnessed in developing economies[1] [3, 4]. In this part of the world, the urban population is expected to double in the next 40 years, from 2.6 to 5.3 billion people [2] (Fig. 1). In general, urbanization is found to influence the world GHG emissions due to the concentration of consumption, labour force and production in cities and the consequent transformation of the structure of the economy, from low-energy agriculture to high-energy intensity production of specialized commodities. Cities concentrate population and economic affluence (at least for some urban inhabitants), which is associated with higher consumption and therefore with a need of constant supply of energy and resources, often from very long distances. Urbanization also tends to increase motorized transport, which is the second energy-consuming urban sector; it implies the construction and maintenance of roads, bridges, office buildings, sewage networks and power plants, all of which require high-energy inputs [2].

Cities are also heat islands—normally warmer than the surrounding areas—and this leads to higher GHG emissions, since more energy is used for air conditioning. The heat island effect is usually higher in large urban centres due to high-density, little open space and reduced natural ventilation. Impervious surfaces, such as roads, buildings, and other constructed areas, absorb and retain solar irradiation, while the displacement of trees reduces natural cooling effects of shading [2, 5].

[1]We apply the United Nations definition of more and less developed regions and least developed countries (http://www.un.org/en/development/desa/policy/cdp/ldc/ldc_definitions.shtml)

At global level, we know that energy consumption is the largest contributor to CO_2 emissions, which in turn is the biggest contributor in GHG emissions. The same is applicable to cities: their contribution to global warming or their '**carbon footprint**' (kg or ton of produced CO_2) is attributed mainly to combustion processes for the production of energy, with fossil fuels holding the greatest share of the urban energy mix [6]. Urban inventories encapsulate several components of the global warming potential, expressed in carbon dioxide equivalents: electricity, heating and industrial fuels, industrial processes, ground transportation, aviation, marine, and waste management. The dominant consumer of energy are the residential and commercial sectors (for light, heating and cooling). According to UNEP, these sectors account for 40 % of the annual energy consumption, 20 % of the annual water usage, and generate up to 30 % of all energy-related greenhouse gas (GHG) emissions. Residential and commercial buildings consume approximately 60 % of the world's electricity. The second largest consumer of urban energy is transportation, accounting on average between 25 and 38 % of the total consumption [6]. According to UN Habitat, the mobility in cities is on a constant ascending slope: by 2005, approximately 7.5 billion trips were made in cities worldwide each day; in 2050, there may be three to four times as many passenger-kilometres travelled as in the year 2000 [7].

In order to have a complete picture of the urban contribution under climate change, it is important to look beyond these general figures and discern the huge disparities among cities in terms of energy consumption and GHG emissions. This uneven distribution has been associated to several traits of the urban economy and morphology: the geographical location of the city, whether its economy is service or industrial oriented, the population density and distribution of settlement, the level of wealth and the access to technology for waste transformation, all these elements have an impact on the overall accounting of GHG emissions. Thus, the type of climate and the corresponding hot and cold days that each city enjoys influences the energy consumed on heating or cooling. The type of energy a city has access to is also relevant: for instance, if a city is close to hydropower or nuclear from nearby sources it tends to produce less emissions than the cities that have easy access to coal seams [8].

Urban form is an important variable whereby urban density is inversely correlated to its energy use and emissions. Low-density areas such as rural regions or suburbs have less traffic congestion and usually benefit from more open and green space thus reducing the urban heat island effect (ergo are characterized by lower energy consumption related to air conditioning). Overall however, they tend to consume more energy due to higher number of transport miles, longer electricity transmission and the inability to rely on district heating and combined heat and power systems. In turn, compact cities have the potential to reduce the private transport, and are more suitable for bicycling and walking [9]. For example, density is found to be the element that accounts for the different emissions per capita from ground transportation between Denver, one of the less dense cities in the USA is 6.31 t e CO_2 and Barcelona, more compact, where the same emissions are 0.77 t e CO_2 [1]. However, increasing density alone can have perverted effects: between tall buildings there is less ventilation since air mixes less and instead it passes over the tops of the buildings. The result is that although there may be fewer emissions due to less traffic, the air quality decreases and the heat island maintains its intensity because of emission concentrations between buildings.

Economic affluence of the urban residents is another key variable. Studies have found a clear direct relationship between income and carbon footprint: to the extent to which more affluent people prefer to live in cities, these urban inhabitants can afford to consume more energy services and therefore pollute more. However, there are exceptions to the rule: when comparing London to Mexico City, even if the first has four times the per capita income of the latter, London has a much lower carbon footprint—almost 40 % less—than Mexico City. While in Mexico City car emissions are still responsible of 35 % of the total, in London the congestion pricing had positive effects thanks to the modal shift to mass transit. Other interventions in the building industry—setting energy standards and incentives for combined heat and power—have also helped London balance to a certain extent the negative environmental impact of its high purchase power [9].

The income per capita variable has generated considerable scholarly debate about the adequate methodology to measure the urban carbon footprint, more specifically among those who defend inventories based on consumption and those who prefer the production-based inventories. The debate revolves around the following question: if the affluent inhabitants of London consume goods that have been produced elsewhere in the world (up to 80 %), to whom should the emissions associated with the production of these goods be attributed: the Londoners or the residents of the places where these goods are produced [1, 10]. In the next section we discuss this and we provide a brief overview of several methodologies for elaborating emissions inventories, a key element for designing and implementing the urban climate action.

Measuring the Carbon and Environmental Footprint of Cities

Determining the carbon footprint of a city or a community implies several methodological choices, which have traditionally generated scholarly disagreement. For instance, the contribution of a city to global GHG emissions can be significantly different if its inventory accounts only for direct emissions (i.e. produced by operations occurring within local boundaries by activities such as transport, commercial or industrial processes, treatment of waste) if it also includes indirect emissions (i.e. those produced outside city limits, for example, by business related air travel). Emissions from electricity can be a relevant example here, since electricity is normally produced outside the city borders but largely consumed within them. Likewise, transportation can weight differently in the balance if the emissions generated by the commuters are considered or not in the measurement [8, 11, 12].

As mentioned in the previous section, another important premise that can determine significant variations in the results of a measurement resides in the difference between production-based emissions (the emissions are associated to the goods and services produced in a given city) or consumption-based ones, where emissions are assigned to the city where the people consuming those goods live. In this sense, the distribution of energy end-use can be very different in the developed and developing world: while in Europe the urban economies have shifted from industry to services, in many developing economies industry still represents the largest share of the economic activity of cities, with Shanghai reaching over 80 % of energy for industry [6] making this city a large contributor to GHG emissions, if a production-based methodology is used. However, if we take into consideration that 33 % of the total production of Shanghai industry is destined for exports [10], a consumption-based inventory would alleviate considerably the responsibility of the city to climate change.

At global level, the United Nations Framework Convention on Climate Change (UNFCCC) requires member countries to identify and report regularly on their GHG emissions. In this sense, the Intergovernmental Panel on Climate Change [13] has provided a methodology that aims to help countries measure their emissions in four sectors: energy; industrial processes and product use, agriculture, forestry and other land use; waste. Until recently there was no such comprehensive framework available at the urban scale [8]. In 2014, the ICLEI issued the first **Global Protocol for Community-Scale Greenhouse Gas Emission Inventories (GPC)**. It provides the most comprehensive greenhouse gas accounting and reporting framework for cities available so far. Pilot application of the protocol confirms the tool sets out clear rules for accounting but also points out the practical difficulties to obtain sufficiently disaggregated data. One identified weakness of GPC is that it is predominantly a production-based inventory, whereas community level action needs an inventory that also took into consideration the emission sources that can actually be influenced by that community [14] (Fig. 2).

In a context of a rising public awareness about the impact of modern lifestyles and consumption patterns on our natural environment and climate change, measuring the environmental footprint of urban economic processes or products consumed by city inhabitants has received increased attention. One effective way to assess this is to determine how much **embodied energy** (the total amount of non-renewable energy used in production) is required. Taking for instance the food consumed in cities, apart from the energy used to produce it, we know it is being transported further than ever before, often by air between countries on opposite sides of the world. Beyond the carbon footprint, increasing ‘food miles’ has other side effects such as increasing road congestion, noise and stress [15]. In many industrialized cities, there is a growing support to bringing production of food closer to or even into the city via urban agriculture.

A more comprehensive way to track the influence of cities on the environment is the **Ecological Footprint** [16] which measures a city or region’s demand for natural capital as compared with the amount of natural capital actually available, expressed in the productive area (gha) required to provide the renewable resources humanity

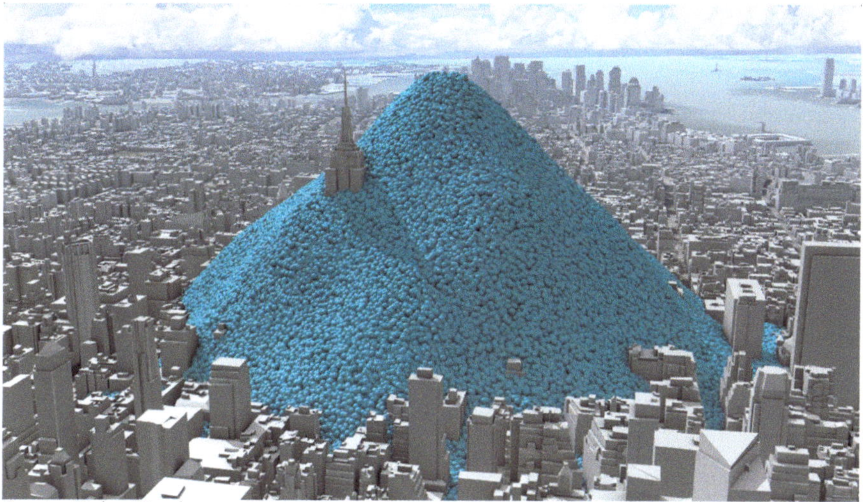
Source - Carbon Visuals

Fig. 2 Visual representation of New York CO_2 emissions in one day

is using and to absorb its waste.[2] The carbon footprint is incorporated in this measurement and represents its largest (54 %) and fastest growing component (Fig. 3).

Additionally, the Ecological Footprint shows us "how carbon emissions compare and interact with other elements of human demand, such as the pressure on food sources, the quantity of resources required to make the goods we consume, and the amount of land we take out of production when we pave it over to build cities and roads" [16]. Applied to cities, the Ecologic Footprint confirms the aforementioned importance of income on a city's overall sustainability: while density and public transportation options significantly reduce the per capita footprint, increased affluence of city residents correlates with higher consumption and therefore results in a larger ecologic and carbon footprint.

[2]The Global Footprint Network defines this tool like this "The Ecological Footprint measures the *supply* of and *demand* on nature. On the supply side **biocapacity** represents the planet's biologically productive land areas including our forests, pastures, cropland and fisheries. These areas, especially if left unharvested, can also absorb much of the waste we generate, especially our carbon emissions. Biocapacity can then be compared with humanity's *demand* on nature: our **Ecological Footprint.** The Ecological Footprint represents the productive area required to provide the renewable resources humanity is using and to absorb its waste. The productive area currently occupied by human infrastructure is also included in this calculation, since built-up land is not available for resource regeneration."

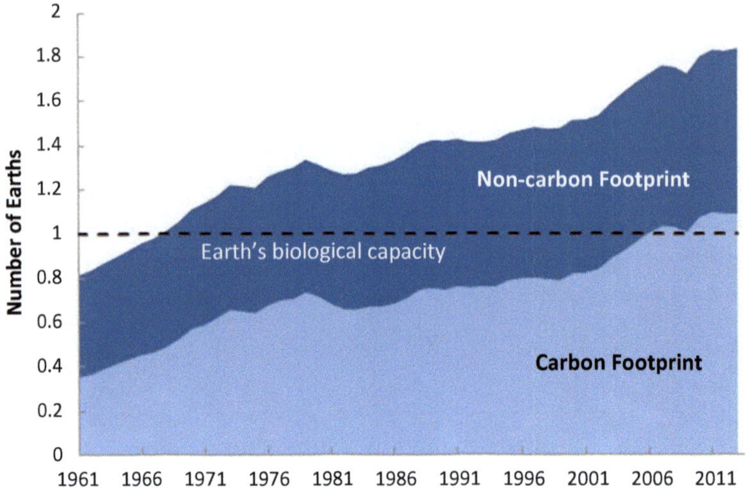

Fig. 3 Ecological and carbon footprints. Source—Global Footprint Network

Cities as Solution or as a Problem

While the scope and methodologies employed for different inventories can be a matter of discrepancy, the overall figures are beyond contestation: cities consume 75 % of natural resources, produce 80 % of global GDP and are home to the majority of world population. Therefore, it is fair to affirm that the global effort for sustainability can be won or lost in the world's cities. Why cities are considered a platform for the battle against climate change and resource depletion? Apart from the reasons already mentioned in this article, "cities are hubs for ideas, commerce, culture, science, productivity, social development and much more. At their best, cities have enabled people to advance socially and economically" [17]. Obviously climate change cannot be fought by municipal administrations alone and it needs strong commitment from all levels of government. However, in this section we focus on a few examples of climate actions that can be taken successfully by cities.

So far, we have shown that urbanization changes the patterns of energy consumption and can produce an increase in emissions due to private transportation, the heat island effect or the concentration of population and industry. Nevertheless, available research shows that it is within the reach of urban administration to curb at least some of the effects of urban morphology or consumption patterns on climate. In this sense the density of urban areas can actually be regarded as an opportunity rather than a problem: policies promoting more energy-efficient forms of housing or the reduction of transportation emissions can have a greater impact in big, compact cities than in less densely built environments [4]. Although the concentration of economic activity and population is expected to generate greater needs for energy supply and therefore increased GHG emissions, recent authors provide a more nuanced perspective: "economies of scale, proximity and agglomeration mean that it is cheaper to provide the infrastructure needed and services needed and to minimize environmental hazards; the concentration of enterprises means that it is less costly to enforce environmental legislation; and the relative proximity of homes and businesses can encourage walking, cycling and the use of mass transport in place of motor vehicles" [8]. Moreover, urban planning can ensure that long-lived infrastructure—such as residential and commercial buildings, water and transport networks—are designed to increase the energy efficiency of the built environment [13]. Other measures include the use of integrated energy production such as combined cooling, heat and power or coding and labelling schemes that aim to reduce the demand for both new and existing buildings. For instance, in the UK, new buildings or major refurbishments are required to pursue at least 10 % GHG emissions reduction by installing on-site generation of electricity from renewable source [9, 11].

Reducing the energy requirements of goods consumed within the city for instance by promoting productive urban landscapes—rooftop gardens, community gardens, etc.—that include urban agriculture is yet another way to adapt to climate change from the municipal level. A city like London could produce about 30 % of all fruit and vegetable requirements of its population within the city boundary and it could achieve this by only using currently abandoned, leftover space. Reductions in 'food-miles' can help curb personal carbon emissions by an average of 950 kg CO_2/year, and create social and health benefits for allotment tenants thanks to daily physical activity, and community interaction [15].

Transport is a sector where cities also have an important say. Sustainable mobility includes several dimensions: energy-efficient and affordable public transport systems; a friendly environment for soft transport modes such as cycling and walking; local transport networks well connected to regional networks; applying congestion charges to discourage private transportation in the city centre, etc. [17]. Additionally, promoting the use of electric vehicles by enabling recharging services is yet an important measure that resides at the urban policy level [6, 11, 17].

2 Conclusions

From this general overview of urban energy consumption and GHG emissions, we can conclude that cities can be seen as both a problem and as a solution for climate change. On the one hand, they concentrate a great share of the current economic activities implicating a high-energy demand; on the other hand, they also provide a level of policy-making that enables direct and concrete action to tackle the negative effects of economic development. As we transition to a more urban future, effective city planning should be increasingly linked to climate change adaptation: a more compact urban growth that avoids heat island effects, more areas devoted or converted to urban agriculture, more mass transit, greater use of cleaner energy for transport and buildings, more sustainable consumption at urban level can all play a positive role. Nevertheless, in order to promote a low-carbon future, the development of inventories and tools for analysis of energy use and emissions are an important for cities to guide their decision-making, the allocation of scarce financial resources, to set targets and held institutions to account.

There is already a wide range of tools and methodologies available for cities in this endeavour. Apart from the new Global Protocol for Community-Scale Greenhouse Gas Emission Inventories, Ecological Footprint measurement mentioned in this article is another valuable tool for policy makers, providing information about their city resource metabolism. Finding the right methodology for assessing the urban environmental performance can help attribute a concrete climate responsibility to human settlements, raise the awareness of local policy makers and citizens about environmental sustainability, and push cities to take action. Importantly, it should contribute to drawing our attention to the key variable of climate change: the current unsustainable pattern of consumption of affluent societies. Climate change is a complex challenge and adapting to it successfully demands integrated action across policy and territorial levels, involving national governments worldwide, local authorities, the industry and the urban citizens.

3 Summary

- The majority of the world population now lives in cities and urban centres. The concentration of population and economic activities results in a high demand for energy and large GHG emissions. However, this concentration also represents an opportunity to take effective action at local levels to mitigate climate change. Cities are fertile ground for innovation, science and knowledge that can be put to the service of sustainable patterns of human activities.
- There are huge disparities between cities in terms of energy consumption (overall and in terms of sectors) and it is difficult to assess the exact contribution of each city to climate change. It is therefore important to create and consolidate frameworks of collaboration among and across policy levels to tackle this challenge more effectively.
- Consumption-based methodologies for carbon footprint measurement can contribute to drawing our attention to the key variable of climate change: the current unsustainable pattern of consumption of affluent countries. They can guide both community and individual action.

References

1. OECD: Competitive Cities and Climate Change, Conference Proceedings, Milan, Italy (2008)
2. Reinhard, M., Sunak, Y.: Impacts of urbanization on urban structures and energy demand: what can we learn for urban energy planning and urbanization management. Sust. Cities Soc. **1**(1), 45–53 (2011)
3. Rutter, P., Keirstead, J.: A brief history and the possible future of urban energy systems. Energy Policy. **50**, 72–80 (2012)
4. Satterhwaite, D.: How urban societies can adapt to resource shortage and climate change, Review. Phil. Trans. R. Soc. A. **369**, 1762–1783 (2011)
5. O'Neill, B.C., Dalton, M., Fuchs, R., Jiang, L., Pachauric, S., Zigovad, K.: Global demographic trends and future carbon emissions. Sust. Sci. PNAS. **107**(41), 17521–17526 (2010)
6. UN Economic Commission for Europe: Climate Neutral Cities: how to make cities less energy and carbon intensive and more resilient to climatic challenges (2012)
7. UN Habitat: Energy Consumption in Cities, in State of the World Cities 2008/2009. Harmonious Cities (2009)
8. Dodman, D.: Blaming cities for climate change? An analysis of urban greenhouse gas emissions inventories. Environ. Urban. **21**, 185 (2009)
9. Sovacool, B., Brown, M.: Twelve metropolitan carbon footprints: a preliminary comparative global assessment. Energy Policy. **38**, 4856–4869 (2010)
10. Brander, M., Carstairs, S., Topp, C.F.E.: Global protocol for community scale greenhouse gas emissions: a trial application in the West Highlands of Scotland. Greenhouse Gas Measurement and Management. **3**(3–4), 149–165 (2013)
11. Hoornweg, D., Lugar, S., Trejos Gómez, C.L.: Cities and greenhouse gas emissions: moving forward. Environ. Urban. **23**, 207 (2011)
12. Kennedy, C., Steinberger, J., Gasson, B., Havránek, M., Pataki, D., Phdungsip, A., Ramaswami, A., Villalba Mendez, G.: Greenhouse gas emissions from global cities. Environ. Sci. Technol. **43**, 7297–7302 (2009)
13. van den Hoeven, M.: Urban energy policy design. Sust. Cities. www.sustainablecities2012.com (2012)
14. Jollands N: Cities and energy, OECD. Competitive Cities and Climate Change, Conference Proceedings, Milan, Italy (2008)
15. Bohn, K., Viljoen, A.: The edible city: envisioning the Continuous Productive Urban Landscape (CPUL). Field J. **4**(1), 149 161 http://www.field-journal.org/uploads/file/2011%20Volume%204/10%20The%20Edible%20City%20Katrin%20Bohn%20and%20Andre%20Viljoen.pdf
16. Global Footprint Network, Footprint for Cities: http://www.footprintnetwork.org/en/index.php/GFN/page/footprint_for_cities/
17. UN, Sustainable Development Goals: http://www.undp.org/content/undp/en/home/sdgoverview/post-2015-development-agenda.html
18. ICLEI, Global Protocol for Community-Scale Greenhouse Gas Emission Inventories: http://www.iclei.org/our-activities/our-agendas/low-carbon-city/gpc.html

Information Sources

C40cities—Climate Leadership Group: http://www.c40.org/
Governments for Local Sustainability (ICLEI): http://www.iclei.org/
UN Habitat: http://www.unhabitat.org/
UN Sustainable Cities: http://www.un.org/en/sustainablefuture/energy.html

UN Resource Efficient Cities: http://www.unep.org/resourceefficiency/Policy/ResourceEfficientCities/tabid/55541/Default.aspx
International Energy Agency (Climate Change section): http://www.iea.org/topics/climatechange/
World Mayors Council on Climate Change: http://www.worldmayorscouncil.org/
The Global Network of Cities, Local and Regional Governments (Climate change action section): http://www.uclg.org/en/issues/climate-change

Lucia Alexandra Popartan is an associated researcher in the research group Multilevel Governance in the European Union, Autonoma University of Barcelona (UAB). She is a manager of European projects at the Institute for the Environment at Universitat de Girona (UdG) and member of the LEQUIA research group at UdG.

Francesc Morata was a Professor of Political Science at the Autonomous University of Barcelona (UAB). He was a Jean Monnet Chair Ad Personam in European Integration, and the director of the research group on Multilevel Governance in the European Union at UAB.

Low Carbon Urban Design: Potentials and Opportunities

Edwin H.W. Chan, Sheila Conejos, and Michael Wang

Abstract The importance of urban sustainability has led to the acceptance of the core concept and principles of low carbon urban design among researchers and professionals worldwide. The challenging issue of climate change urgently calls forth cities working towards a sustainable low carbon future. Cities can work together to seek solutions for carbon neutrality to deal with climate change and build the foundation for urban sustainability. This chapter introduces the concept of low carbon urban design, discusses the potentials and opportunities of a low carbon urban development and defines its role in making cities achieve carbon neutrality. Examples of low carbon urban design initiatives in Australia, China and Hong Kong are deliberated. Further research on low carbon urban design policies, strategies and technologies as well as the creation of a standard low carbon urban design indicators list that is applicable to existing and new cities is the way forward.

Keywords Low carbon urban design • Urban sustainability • Indicators • Australia • China • Hong Kong

Key Terms

- Urban design involves the design of buildings, public spaces, landscapes, transport systems, services and amenities.
- Low carbon urban design aids the design of livable cities with high living quality that considers the different aspects of urban development such as technology, energy efficiency, health, social and economic.

E.H.W. Chan (✉)
Research Institute of Sustainable Development, and Building and Real Estate Department, Hong Kong Polytechnic University, Hung Hom, Kowloon, Hong Kong
e-mail: edwin.chan@polyu.edu.hk

S. Conejos
Department of Building, School of Design and Environment, National University of Singapore, 4 Architecture Drive, 117566 Singapore, Singapore

M. Wang
Building and Real Estate Department, Hong Kong Polytechnic University, Hong Hom, Kowloon, Hong Kong

S. Dhakal, M. Ruth (eds.), *Creating Low Carbon Cities*,
DOI 10.1007/978-3-319-49730-3_8

1 Low Carbon Cities and Urban Sustainability

As major drivers of global environmental change, cities have become the focus of sustainability. It was reported that 70 % of the world's greenhouse gas emissions are accounted from cities [1] and 70 % of the world's population will live in cities by 2050 [2]. Therefore, cities play a vital role in combating climate change issues. The challenging issue of climate change calls for an urgent need to reduce carbon emissions and plan for a sustainable low carbon future [3, 4] among cities in the world. Furthermore, planning and policies for reducing carbon emissions and stabilizing greenhouse gases must be established.

The United Nations Environment Program [4, 5] defined climate neutrality as "living in a way which produces no net greenhouse gas (GHG) emissions" which can "be achieved by reducing your own GHG emissions as much as possible and using carbon offsets to neutralize the remaining emissions." Indeed, if cities work together to seek solutions for carbon neutrality, then climate change can be dealt with to build the foundation for urban sustainability. This chapter introduces the concept of low carbon urban design, discusses the potentials and opportunities of a low carbon urban development and defines its role in making cities achieve carbon neutrality.

The concept of a low carbon city reflects on the national aspiration to create "low carbon economies" which systematically incorporate mitigation and adaptation measures to enable cities to reduce and respond to climate change through a well-planned and designed urban environment [6]. To cope with climate change, the concept of low carbon cities has emerged and has been incorporated in urban planning and design practices.

A growing awareness of the importance of urban sustainability has led to the acceptance of the core concept and principles of low carbon urban design among researchers and professionals around the world. This is a new terminology in academic articles that has been combined with or linked into a wide variety of relevant concepts such as eco-city, sustainable community and green buildings. Through the comprehensive and continuous application of low carbon urban design practices, cities and urbanized areas could become more environmentally friendly and develop towards a low carbon future.

2 Low Carbon Urban Design Potentials

There are many different emphases tendered for the definition of urban design. This chapter prefers the version in the Planning Policy Guidance Note 1 of the Department of the Environment, Transport and the Regions of the UK, which defines urban design as "the relationship between different buildings; the relationships between buildings and the streets, squares, parks, waterways and other spaces which make up the public realm; the relationship of one part of a village, town or city with other parts; patterns of movement and activity which are thereby established; in short, the complex relationship between all the elements of built and unbuilt space" [7]. Urban design involves the design of buildings, public spaces, landscapes, transport systems, services and amenities. Each of these components plays significant roles in contributing to carbon emissions of a city.

Urban design, in creating sustainable urban development, should balance the economic growth, care of our environment and social progress. For low carbon urban design, we focus on achieving low carbon emissions but still strive to maintain the balance required for the three pillars of sustainable development. The role of low carbon urban design is to aid the design of livable cities with high living quality with consideration of different aspects of urban development such as technology, energy efficiency, health, and desirable social and economic attributes. The promotion of low carbon urban design will improve the social and ecological health of cities and its buildings.

As one of the important components in urban planning, urban design supports low carbon urban development through the design and management of functional, attractive and sustainable places for people. It has eight specific elements, which include land use; building form and massing; circulation and parking; open space; pedestrian ways; activity support; signage and preservation. Low carbon urban design adopts modern and environmentally friendly approaches, which include the following characteristics [8–10]:

- Mobility and connectivity—extensive use of public transport systems and pedestrianized areas, integrated zones for residential, industrial and commercial use and their interlinkages.
- Compact, mixed use development and optimal building density—vertical as well as lateral structures.
- Greater adoption of renewable energy and resource efficiency.
- Urban water demand management—an integrated approach for safeguarding of supply sources and maintaining efficient delivery and disposal systems.
- Urban greenery and vegetation—large park systems and open spaces with extensive greenery and natural ventilation systems.
- Smart growth and green infrastructures-reduced urban sprawl and construction of green buildings or green retrofitting.
- People friendly environment—such as public spaces, walking and bicycling areas.
- Community focus and community delivery—community level planning approach and delivery.
- Low carbon communities living programmes—community based low carbon approaches that encourage people to take practical action to reduce their carbon footprint.

Apart from these urban design and planning guidelines and indicators to ensure a low carbon future, there is the emphasis to reduce energy and GHG emissions associated with buildings that consume 40 % of the world's energy. Thus, green buildings, [11] adaptive reuse [4] and future "adaptivity" of new buildings [12] has a significant role in global climate protection and emission reduction. In the future, cities will focus on smart solutions such as Smart grids, water systems, public safety and intelligent buildings [13]. To address climate change issues, buildings will be designed to be intelligent with greater flexibility and wireless sensor networks, holistic energy management systems, web enabled services, and installation with, smart appliances and BIM [13].

Existing Indicators for Urban Sustainability and Low Carbon Performance

Achieving a low carbon city with the right results needs sustainability measurements. A range of sustainability indicators endorsed by international and regional organizations as well as green building rating systems, protocols, guidelines and standards has been developed to evaluate and benchmark levels of sustainability in countries and cities [14].

For example, Chan and Lee [15, 16] suggested six significant design criteria that could be integrated for urban environment sustainability such as Land Use Planning, Quality of Life, Conservation and Preservation, Integrated Design, Provision of Welfare Facilities and the Conservation of Existing Properties. The framework of sustainable urban renewal model [17] as shown in Fig. 1 also ventures to assess the economic, environmental and social sustainability of urban renewal projects.

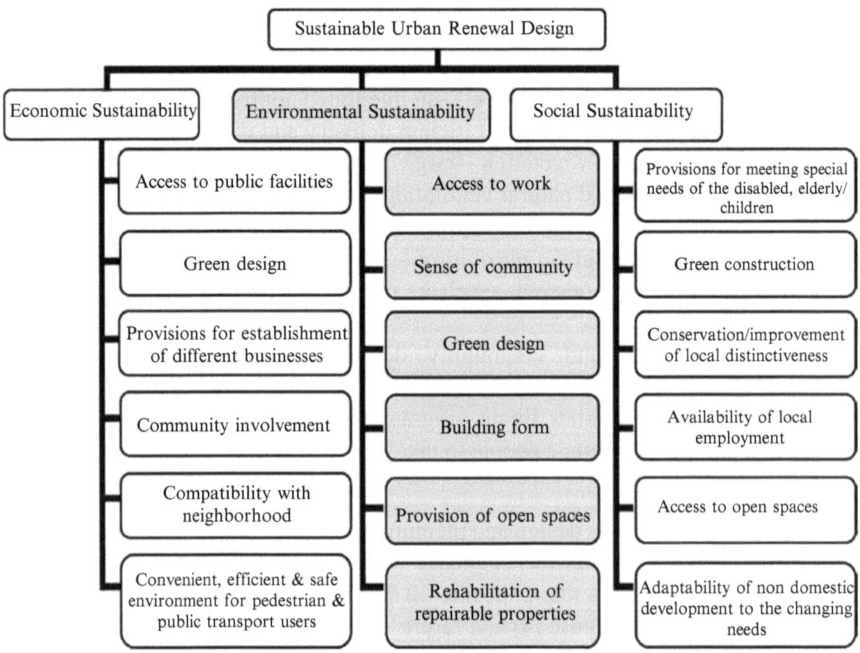

Fig. 1 Sustainable urban renewal model [17]

Another development for measuring performance is the Sustainable Systems Integrated Model (SSIM), which measures the sustainability and low carbon performance of development projects as well as helping developers make strategic decisions based on what they will get for a certain sum of investment [9]. There is also the Low Carbon Future Cities project that "aims to develop an integrated urban low carbon adaptation and circular economy strategy to harness the potential for CO_2 mitigation in urban areas by engaging cities and stakeholders in China and Germany in an integrated approach" [3]. There are a number of research studies conducted to establish measurements, which identify numerous design criteria that will meet the three pillars of sustainability such as economic, environmental and social sustainability. Although still lacking is the development of a universal and standardized list of design indicators for low carbon urban development.

Potential Directions: Emergent and Future Technologies

A number of researches have revealed great potentials of advanced approaches and frontier technologies for reducing carbon emissions in modern cities. In the field of urban design, multi-functional urban structure, mixed land use and compact urban form are recognized as three important design concepts that will reduce internal transportation demands and relevant carbon emissions significantly [15]. Numerous new technologies, such as solar refrigeration and cooling technology, natural lighting, and ecological water recycling systems also provide considerable carbon reduction potentials for cities owing to their higher efficiency and lower energy consumption capabilities [11].

In an attempt to enhance potentials for a low carbon future of cities, effective technologies, approaches and strategies are integrated and implemented in Australia and China through the application of urban design guidelines and frameworks [18, 19]. In recent years, various guidelines are proposed by researchers and governments to ensure urban designs that are executed in efficient and effective ways and also match the specific conditions and requirements of local cities. In China, a set of urban design guidelines entitled "Low Carbon City: Principles and Practices for China's Next Generation of Growth" was formulated with eight design principles as described below [19]:

1. Develop Neighbourhoods that Promote Walking
2. Prioritize Bicycle Networks
3. Create Dense Networks of Streets and Paths
4. Support High-Quality Transit
5. Zone for Mixed-Use Neighbourhoods
6. Match Density to Transit Capacity
7. Create Compact Regions with Short Commutes
8. Increase Mobility by Regulating Parking and Road Use

Table 1 An urban design protocol for Australian cities [18]

Goal	Framework	Design principles
Creates productive, sustainable and livable places for people through leadership and the integration of design excellence.	Productivity and sustainability	• Enhancing: enhances the local economy, environment and community • Connected: connects physically and socially • Diverse: diversity of options and experiences • Enduring: sustainable, enduring and resilient
	Livability	• Comfortable: comfortable and welcoming • Vibrant: vibrant, with people around • Safe: feels safe • Walkable: enjoyable, easy to walk and bicycle around
	Leadership and design excellence	• Context: works within the planning, physical and social context • Engagement: engages with relevant stakeholders • Excellence: fosters excellence, innovation and leadership • Custodianship: considers custodianship and maintenance over time

The urban design guidelines implemented for Australian cities as shown in Table 1. Others, such as the "By Design—Urban design in the planning system: towards better practice" published UK's DETR, [7] provide graphic guidelines to strongly promote the need of urban design with raising standards for sustainable urban development.

3 Low Carbon Urban Design Opportunities

A number of sustainable low carbon city developments are ongoing around the world. Actually, low carbon communities are not only evident in the USA and Europe, but most low carbon cities are being developed in Asia like China and India [6] and the rest of the world. A good example is Masdar in the United Arab Emirates, which is considered as a brand new zero-carbon city in the Middle East [6]. In the following, examples of low carbon urban design initiatives in China and Hong Kong are discussed.

Low Carbon City Initiatives in China

Generally, there are two major opportunities of low carbon urban design; one is "new town development", and another is "urban renewal". Nowadays in the world, most of new town development projects are located in developing countries, such as

China, India, and Brazil, where urbanization is moving forward rapidly. In China, the development of low carbon cities is widely recognized as a crucial and also effective strategy for achieving the country's goal of reducing carbon emissions per unit of GDP by 40–45 % by the year of 2020, based on the level of 2005 [20, 21]. New town developments offer a great opportunity for building low-carbon new cities since low carbon urban design elements can be taken into consideration at the beginning of planning phase. Table 2 demonstrates two examples of new town development with low carbon initiatives implemented in China [20, 21].

For the rest of the world, a widely applicable but more challenging opportunity for promoting low carbon urban development is through urban renewal projects, in which urban design is employed as a basis for reshaping a city in renewal process within its existing urban environment. Some advanced low carbon technologies or measures are utilized comprehensively, including but not limited to mixed land use, low carbon transportation system, eco-system, waste-recycling system and building energy efficient retrofits. Since every city needs to respond to contemporary city life and will be undergoing urban reconstruction sooner or later both in developed and developing countries, urban renewal will present important opportunities for facilitating cities to achieve low carbon development goals in the future. For example, in Hong Kong, during the period of economic boom from 1970s, thousands of buildings and city facilities were built to answer demands from its rapid increasing population. As time goes by, a great proportion of these buildings becomes obsolete, and needs refurbishment or reconstruction now. Faced with this situation, Hong Kong is encountering huge pressures of building retrofitting and urban renewal. From another perspective, this is a great opportunity for exerting low carbon urban design to transform Hong Kong to become a low carbon city, and enhance its urban sustainability. One exemplary case enabling low carbon city initiatives in Hong Kong is the "Kai Tak Development" to be discussed in the following section.

Application of Low Carbon Urban Design in Hong Kong: The Kai Tak Development

To create a high quality, sustainable built environment in Hong Kong, due consideration is given to urban design concepts and principles in the planning and development process. The pursuit for low carbon urban planning in Hong Kong came from the government's pledge to reduce its carbon intensity by 50–60 % below the 2005 baseline by the year 2020. Since urban design is essential for a compact and dynamic city like Hong Kong, a series of planning policies are established to reinforce its low carbon characteristics.

There are two significant green initiatives that act as pioneer projects for sustainable urban development in Hong Kong. One is the construction of its first Zero Carbon Building (ZCB) in 2012, which showcases up-to-date green design

Table 2 Case studies of two low carbon new town developments in China (extracted from various sources [20, 21])

Project	Location	Area	Expected residents	Design Principles	Low-carbon initiatives	Status
Sino-Singapore Tianjin Eco-City	Tianjin Binhai New Area (40 km from Tianjin city centre), China	31.23 km^2 eco-city project in the Tianjin-Binhai New Area (TBNA)	350,000	• Group-style layout • Public transportation oriented • Mixed land use • Priority of public interests	• Renewable energy: be introduced in various forms, excepted to meet the goal KPI of "20 % renewable energy usage" in its low-carbon vision. • Green buildings: all the buildings are designed and built in line with the Green Building Evaluation Standards (GBES), excepted to save at least 70 % of energy demand, with conventional build environment. • Low-carbon transport: be expected by 2030, 90 % of the daily trips of city residents will be green, by using of electrical private vehicles, and public transport, like light rail transit (LRT), tram and electrical buses, cycling or walking. • Smart grid: enable a new lifestyle intelligent home energy management system, digitalized civic services enables a new lifestyle towards energy efficient and sustainable society.	Launched in January 2008; 12 km^2 of infrastructure have been built by 2015; permanent resident population exceeds 30,000 in 2015

Tangshan Bay Eco-City	Caofeidian new district, Tangshan, Hebei Province, China	30 km^2 (first phase, with an initial area of 12 km^2)	400,000–500,000	Ten principles: • Focus on people • Resource conservation • Green buildings • City security • Recycling economy • Green transportation • Renewable energy • Diversity lifestyle • Cultural integration • High efficient public utilities	• Renewable energy: including wind power, solar power, waste to energy, and tidal energy. • Eco-recycling technologies: sewage treatment, combined heat and power (CHP), neutral water recycling. • Resource management centre: to integrate eco-recycling system, manage energy, water and waste systems, provide technical support on new energy application. • Mixed land use. • Green transportation. • Green land systems. • Public service centre. • Zero carbon emission	Launched in March 2009; transportation systems, sewage treatment plant, central Heating Station, and other infrastructures have been built; one college campus has been put into operation and another one is under construction by 2015

Fig. 2 Kai Tak Development Project (2014)

and low carbon technologies to promote the concept of sustainable living to the community [22]. The other is the Kai Tak Development project [23], which is an example of a sustainable urban development initiative which incorporates low carbon urban design features. The latter is discussed in details as a low carbon urban design example in Hong Kong as shown in Fig. 2.

Kai Tak was once the location of the Hong Kong International Airport. Since the airport was moved to Chek Lap Kok in 1998, Kai Tak became the largest land available for development in Hong Kong. Considered to be Asia's World City and known for its finance and logistics achievements, Hong Kong aspires to become an attractive destination for tourism, sports events and improved quality of life for its local population. These objectives became the foundation for the Kai Tak Development (KTD) project. As a highly complex development project spans over 320 ha, KTD covers the ex-airport site and the adjoining hinterland districts of Kowloon City, Wong Tai Sin and Kwun Tong. KTD is planned to be a Heritage, Green, Sports and Tourism Hub in Hong Kong based on five salient urban design guidelines that endorse sports-oriented, people-oriented, sustainable, environmental-friendly and distinguished urban form design with some of its key features [23] highlighted as follows:

- Sports-Oriented—A modern Multi-Purpose Stadium Complex is provided as the anchor
- People-Oriented—The waterfront areas are reserved mainly for public enjoyment as parks or promenades with convenient and comfortable pedestrian connections
- Sustainable—Mix-use planning combines with the sports and leisure activity nodes to ensure vibrancy
- Environment-Friendly—Provide solutions to the water pollution and soil contamination problems. Reserve land for roadside greening and district cooling system, and planning for mass transit
- Distinguished Urban Form—Attractive urban form is based on a vision of "Rediscovering the Runway—Taking Off to the Future: A New Harbour-front, City of Heritage, Green, Sports & Tourism".

A range of low carbon features in the approved plan of Kai Tak Development are listed in Table 3.

Table 3 Identified low carbon features in Kai Tak Development

Key principles of low carbon urban design [19]	Low carbon features in Kai Tak Development
1. Develop neighbourhoods that promote walking	Numbers of connective facilities links Kai Tak with surrounding neighbourhoods Open space and waterfront besides the Victoria Harbour Auto-free central plaza and outdoor plaza
2. Prioritize bicycle networks	Bicycle networks around the boundary and links residential area with the great stadium complex
3. Create dense networks of streets and paths	Subject to further planning
4. Support high quality transit	Various transport infrastructure including MTR (the Shatin to Central Link), Central Kowloon Route, Trunk Road T2, and Tseung Kwan O—Lam Tin Tunnel Convenient road network
5. Zone for mixed-use neighbourhoods	Ten types of land use will fulfil various purposes, including commercial, residential, comprehensive development area, open space, government, institution or community, etc.
6. Match density to transit capacity	Kai Tak metro station Kai Tak bus terminus Multiple modes of transportation
7. Create compact regions with short commutes	Zero reclamation Most of commercial and residential buildings locate at the north area Be available for 86,000 peoples to live and 79,600 peoples to work
8. Increase mobility by regulating parking and road use	Subject to further planning

(Source: Kai Tak Development, Civil Engineering and Development Department, HKSAR [23])

4 Future Research

With the rising concern about climate change and global warming, the development of sustainable low carbon urban environments is gaining emphasis globally. There is room for developing a sustainability consciousness in the urban design process in order to achieve a healthier and cleaner built environment that is also productive by making efficient and effective use of its resources. Recent developments to address climate change are geared towards using ICTs to achieve new and alternative ways to deliver services through transformative solutions that support low carbon urban living such as smart motor systems, smart logistics, smart buildings and smart grid, fiber optic cables, mobile network radio base stations and servers [13, 24].

There is also the trend for Strategic Energy Technologies Information System (SETIS) that is an integrated approach for information exchange on energy technologies and capacities for innovation [25]. In addition, there is the challenge of shifting to an information infrastructure development, which is considered "most energy efficient since it provides connectivity and allows information to flow at the speed of light around the planet" [24]. Lastly, further research on low carbon economy should involve different sectors and industries such as ICTs and chemical industries.

In regard to low carbon urban development, the way forward should be the creation of a standardized list of low carbon urban design indicators that is applicable to existing and new cities. Future research on low carbon related strategies such as energy efficient building regulations, green adaptive reuse building rating tools, green technology, sustainable construction and low carbon urban renewal or urban regeneration developments could cultivate a culture with appropriate institutional responses that embraces environmental, ecological, social, physical and political sustainability agenda [26].

Summary

- Low carbon urban design could be better utilized as the process for designing and managing various components of cities that systematically integrates climate change mitigation and adaptation measures.
- There is a potential to build international consensus for a standardized list of low carbon urban design indicators to guide the development of new and existing cities.
- Low carbon urban renewal developments will be the future trend to trim and reshape existing cities to achieve low carbon urban environments.
- In achieving a low carbon future, further research on low carbon urban design policies, strategies and technologies is essential.

Acknowledgement This paper is prepared with the support of research grants from CII-HK (A/C No. 5-ZJF5) and Research Institute of Sustainable Urban Development (A/C No. 1-ZVCJ) of The Hong Kong Polytechnic University.

References

1. Satterwhaite, D.: Cities contribution to global warming. Environ. Urban. **20**(2), 539–549 (2008)
2. World Health Organization (WHO): Global Health Observatory (GHO) Data. www.who.int/gho/urban_health/situation_trends/urban_population_growth_text/en/
3. Venjakob J, Schneider C: Integrated city strategy for CO_2 emission reduction, resource efficiency and climate resilience. Low Carbon Future Cities Report, Wuppertal (2013)
4. Yung, E.H.K., Chan, E.H.W.: Implementation challenges to the adaptive reuse of heritage buildings: towards the goals of sustainable, low carbon cities. Habitat Int. **36**(3), 352–361 (2012)
5. UNEP: A Case for Climate Neutrality—Case Studies on Moving Towards a Low Carbon Economy UNEP, Nairobi, Kenya, p. 52 (2009).
6. Joss, S., Tomozeiu, D., Cowley, R.: Eco-Cities—A Global Survey 2011. University of Westminster International Eco-Cities Initiative, UK (2011)
7. Department of the Environment, Transport and the Regions (DETR): Urban Design in the Planning System: Towards Better Practice. www.designcouncil.org.uk/sites/default/files/asset/document/by-design.pdf
8. Atkins: Planning and Design Handbook for Integrated City Development: A Holistic Approach to a Low Carbon and Resource Efficient Future. www.atkinsglobal.com/~/media/Files/A/Atkins-Corporate/group/sectors-documents/urban-development/LCC_Report_March%202013%20LowRes.pdf
9. Tang: Sustainable Systems Integrated Model (SSIM): Modeling Techniques for Low Carbon Cities. www.building.com.hk/forum/2011_0719modeling.pdf
10. Low Carbon Hub: Developing Low Carbon Communities. www.lowcarbonhub.org/wp-content/uploads/2012/04/Low-Carbon-Communities-Handbook-section-2.pdf
11. UNEP: Building for the Future: A United Nations Showcase in Nairobi. UNEP, Nairobi, Kenya (2011)
12. Conejos, S., Yung, E.H.K., Chan, E.H.W.: Evaluation of urban sustainability and adaptive reuse of built heritage areas: a case study on conservation in Hong Kong's CBD. J. Des. Res. **12**(4), 260–279 (2014)
13. European Commission: ICT for a Low Carbon Economy: Smart Buildings. ec.europa.eu/information_society/activities/sustainable_growth/docs/sb_publications/smartbuildings-ld.pdf
14. Appleby, P.: Integrated Sustainable Design of Buildings. Earthscan LLC, Washington, DC (2011)
15. Chan, E.H.W., Lee, G.K.L.: Design considerations for environmental sustainability in high density development: a case study of Hong Kong. Environ. Dev. Sust. **11**(2), 359–374 (2009)
16. Chan, E.H.W., Lee, G.K.L.: Contribution of urban design to economic sustainability of urban renewal projects in Hong Kong. Sust. Dev. **16**, 353–364 (2008)
17. Chan, E.H.W., Lee, G.K.L.: Indicators for evaluating environmental performance of the Hong Kong urban renewal projects. Facilities. **27**(13/14), 515–530 (2009)
18. Australian Sustainable Built Environment Council (ASBEC): Creating Places for People: An Urban Design Protocol for Australian Cities. urbandesign.org.au
19. Calthrope Associates: Low Carbon City: Principles and Practices for China's Next Generation of Growth. www.calthorpe.com/files/China%20Design%20Manual%20Pamphlet.pdf
20. Zhou N, He G, Williams C: China's Development of Low Carbon Eco-Cities and Associated Indicator Systems. (2012).
21. Gu, C.: Climate Change and Low-Carbon City Planning, pp. 42–45 . Southeast University Press, Nanjing (2013)104–105, 148–153
22. Zero Carbon Building Limited (ZCBL): zcb.hkcic.org/Eng/index.aspx
23. Civil Engineering and Development Department, HKSAR: Kai Tak Development. www.ktd.gov.hk/eng/index.html

24. Pamlin D, Pahlman S, Weidman E: A Five Step Plan for the Low Carbon Urban Development: Understanding and Implementing Low Carbon ICT/Telecom Solutions That Help Economic Development While Reducing Carbon Emissions. www.ericsson.com/res/thecompany/docs/corporate-responsibility/2009/wwf_ericsson_5step_plan.pdf
25. European Union: Strategic Energy Technologies Information System (SETIS) Towards Low Carbon Future. setis.ec.europa.eu/
26. Chan, E.H.W., Choy, L.H.T., Yung, E.H.K.: Current research on low-carbon cities and institutional responses. Habitat Int. **37**, 1–3 (2013)

Edwin H.W. Chan is a Professor and Associate Head (Research) in the Building and Real Estate Department at The Hong Kong Polytechnic University. He is qualified as a chartered architect/surveyor/engineer and a lawyer. He has published over 100 papers in refereed journals and many conference papers and authored books. His research interests include urban policy, social sustainability, urban renewal, green building and dispute resolution.

Sheila Conejos is a Research Fellow in the Department of Building at the National University of Singapore. She has studied in the Philippines, Thailand, Japan and Australia. Her research focus is on building and urban sustainability particularly on green adaptive reuse, green building diagnostics, future adaptive reuse of new buildings, green FM, green maintainability of buildings, heritage conservation and urban regeneration.

Michael Wang is a Ph.D. student from Tongji University and now studying at The Hong Kong Polytechnic University, under the supervision of Prof. Edwin H.W. Chan. His research interests involve low carbon cities, green buildings, and analysis on related influence factors.

Toward Low Carbon Cities: The Chinese Experience

John A. Mathews, Mei-Chih Hu, and Hao Tan

Abstract When surveying the extent to which cities around the world are greening, there is a striking fact about the Chinese experience. It presents both the worst and best facets of the process. The worst is encapsulated by the environmental toll taken on the country, and particularly its cities, in its three decades of high-speed growth—unbreathable air, polluted and undrinkable water, loss of soil, build-up of heavy metal contamination, and many other such problems. On the other hand, and certainly linked to this catalogue of problems, China is also leading the way in terms of solutions. It promotes eco-cities that take sustainability as their development model, and set performance goals in terms of conservation and circulation of resources, utilization of renewable energies, and financing by novel instruments such as green bonds. The issue is: which trend is leading in China?

Keywords China • Low carbon cities • Resource efficiency

Key Terms

1. Greening cities—meaning 'sustainable city' or 'eco-city', where the city is designed with full consideration of environmental impact and with minimization of energy, water and food inputs and heat, pollution and carbon outputs
2. China—the people's Republic of China (PRC), an emerging giant with a total population of more than 1.37 billion in 2015 and urban population accounting for more than half of this
3. Circular Economy—generic term used in China to depict an economy where material flows are designed to circulate at high quality without entering the biosphere

J.A. Mathews (✉)
MGSM, Macquarie University, Sydney, NSW, Australia
e-mail: john.mathews@mgsm.edu.au

M.-C. Hu
Institute of Technology Management, National Tsing Hua University, Hsinchu, Taiwan

H. Tan
Newcastle Business School, University of Newcastle, Callaghan, NSW, Australia

S. Dhakal, M. Ruth (eds.), *Creating Low Carbon Cities*,
DOI 10.1007/978-3-319-49730-3_9

4. Renewable energies—energies derived from resources that are naturally replenished such as sunlight, wind, tides, waves and geothermal sources
5. Sino-Singapore Tianjin Eco-City (SSTEC)—collaborative project between the governments of China and Singapore where the aim is to develop a socially harmonious, environmentally friendly and resource-conserving city for the twenty-first century in China

1 Introduction

When surveying the extent to which cities around the world are greening, there is a striking fact about the Chinese experience. It presents both the worst and best facets of the process. The worst is encapsulated by the environmental toll taken on the country, and particularly its cities, in its three decades of high-speed growth—unbreathable air, polluted and undrinkable water, loss of soil, build-up of heavy metal contamination, and many other such problems. On the other hand, and certainly linked to this catalogue of problems, China is also leading the way in terms of solutions. It promotes eco-cities that take sustainability as their development model, and set performance goals in terms of conservation and circulation of resources, utilization of renewable energies, and financing by novel instruments such as green bonds. The issue is: which trend is leading in China?

In this chapter, we provide an overview of the conflicting trends in China's urbanization, and offer a judgment as to what has been the experience so far, and what might be the results in the near term, up to 2015 (by the end of the period of the 12th Five Year Plan) and to 2020 (the span of the successor 13th Five Year Plan, which was recently released in March 2016). We discuss some of the China eco-city cases such as Qingdao and the Tianjin eco-development zone, and the broader policy environment that is driving the greening of China's cities.

2 China's Urbanization Challenge

China is urbanizing and industrializing at the same time—at a pace unprecedented in history. In the space of just a few decades China has changed, and is changing, from a largely rural to a largely urban population. The figures speak for themselves. China was a largely rural country at the time of the revolution. Then it reached an urbanization level of 20 % by 1980; then 30 % by 1996; then 40 % by 2002 and 50 % by 2011—so by 2012 there were more people living in cities in China than in the countryside (Fig. 1). The urbanization trend is expected to continue. According to the 12th Five Year Plan, China's urbanization level should have reached 54 % by 2015. The latest data released by the National Bureau of Statistics indicate that this

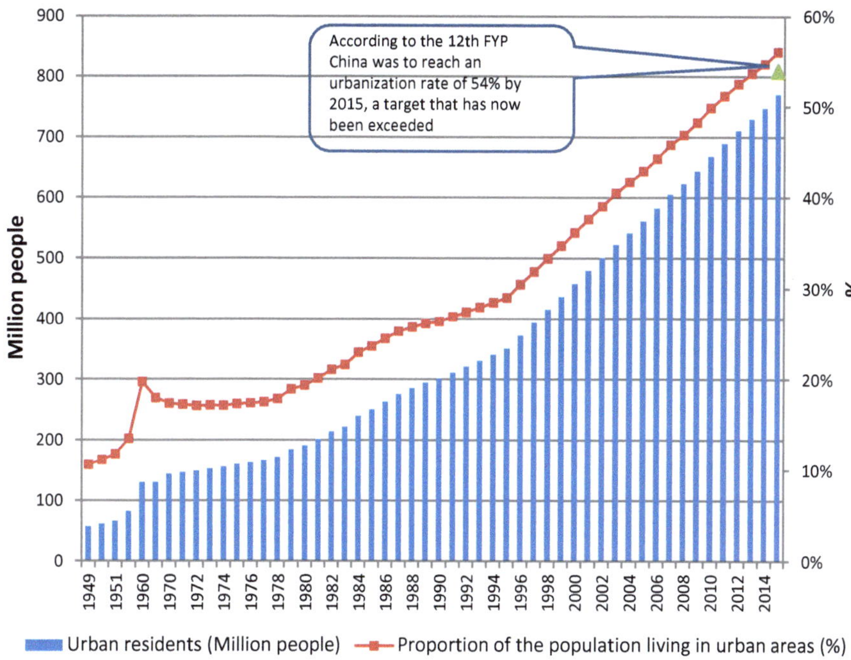

Fig. 1 Urban residents and their proportion in the total population in China: 1949–2015. *Source*: authors based on data from the NBS (2015) [1]. Note: The spike around 1957/58 is not an artefact of statistics, but the result of the decline of the rural population resulting from the three-year 'Great Chinese Famine' of 1958–1960, which was jointly caused by drought and by the policies of the Communist Party of China aimed at making a 'Great Leap Forward'

target has already been exceeded, with China reaching 56 % urbanization level by the end of 2015.[1]

To gain a feel for this frantic pace of urbanization, which drives the greening of China's urban economy, consider the following. Over the decade 2005 to 2015 China raised its number of urbanized residents from 560 to 770 million today. This means an increase of 210 million over the decade—or on average, 21 million newly urbanized people every year. This amounts to building seven new cities of three million inhabitants each, every year—a phenomenal rate of change. No wonder China is viewed as the urbanizing powerhouse of the planet, with a construction and housing industry to match.

Rapid urbanization in the twenty-first century raises critical challenges for China to combat pollution and climate change while utilizing the low carbon cities programme to leapfrog to advanced green construction technologies and practices.[2]

[1]See NBS http://www.stats.gov.cn/english/PressRelease/201602/t20160229_1324019.html

[2]This section is based on Mathews and Tan 2015 [2], pp. 136–137.

Fig. 2 Sustainable KPIs for SSTEC. *Source*: Based on IEK [10]

Compared to its rural living conditions, the modern urban life style tends to lead to higher energy consumption and carbon emissions per capita. Current levels of per capita energy consumption in Chinese cities reflect higher income levels and better quality of life as the city greens its operations. Moreover the new cities represent an opportunity to introduce new green construction and energy-saving technologies [3]. Against this background, it is not surprising that increasing efforts have been made in China to reduce pollution and energy consumption in cities through what is known as the low carbon cities programme.[3]

It was not until 2008 that the concept and term ‘low carbon city’ was introduced in China by the World Wildlife Fund (WWF), a non-government organization [4]. In that year the WWF in collaboration with two municipal governments in China introduced a ‘Low Carbon City Initiative’ which was specifically designed for the context of cities. Among the two participating cities in the initiative, Shanghai was expected to focus on promotion of new eco-buildings and improvement of energy efficiency of existing buildings, and engagement of the public to raise their awareness in energy saving. Another participating city, Baoding, was to facilitate local renewable energy industries as a means toward establishment of low carbon cities.[4] In addition, China’s first carbon trading exchange market was also established in Tianjin in 2008.

[3]See ‘China must take care of its city-dwellers’, by Tom Miller, *Financial Times*, April 14, 2013, http://www.ft.com/intl/cms/s/0/ab9a6376-a358-11e2-ac00-00144feabdc0.html#axzz2mRQqJ011

[4]On these initiatives see Nan Zhou et al. [5] and Li Yu [6].

In 2010, the notion of low carbon cities was picked up at the national level and integrated with the concurrent Circular economy initiatives. In 2010, the National Development and Reform Commission (NDRC), the country's premier policy maker, chose five provinces (Guangdong, Liaoning, Hubei, Shaanxi and Yunnan) and eight cities (Tianjin, Chongqing, Shenzhen, Xiamen, Hangzhou, Nanchang, Guiyang and Baoding) as pioneering cities and provinces for pilots of low carbon initiatives, complemented by Circular economy initiatives.[5] In 2012, the second batch of participating provinces and cities for low carbon pilot projects was announced, including two municipalities (Beijing and Shanghai) and 26 other cities, with emphasis on energy- and water-saving initiatives.[6]

Among those pilot 'low carbon' cities, a number of measures have been introduced, as summarized in Table 1.

We highlight some examples from two cities—Qingdao and Tianjin.

3 Qingdao

Qingdao is a coastal city in Shandong province, well known for its Tsingtao beer and for its hosting the 2008 Olympics sailing competition on its harbour. But Qingdao, with an urban population of 4.6 million in 2015, is also a focus of sustained efforts to create an eco-city based on a Circular Economy. It is a naval base, a seaport and an industrial centre. The city boasts a high level of Japanese and Korean investment. Qingdao High-Tech Industrial Development Zone was approved by the State Council in 1992 and now forms the basis of the city's Circular Economy initiatives.[7]

One of the principal environmental problems faced by Qingdao was the stockpiling of chromic slag, a product of the Qingdao Redstar Chemical Group formed in the manufacture of chromic salts and resulting in a 'mountain' of chromic waste severely contaminating ground water. A solution was found in 2005 with the Qingdao Iron & Steel Group developing a method for using chromic slag as replacement for dolomite (naturally occurring calcium-magnesium carbonate) in the sintering process in iron production. This is a typical 'Circular Economy' solution where a waste is turned into a valuable input into a different industrial process—modelled on natural cycles where 'waste equals food' [9].

This and many more examples of Circular Economy initiatives reveal that national policy, in the form of the Circular Economy Law of 2007, is driving

---

[5]See the 'Notice from the NDRC about carrying out the work of low-carbon provinces, autonomous regions, and cities pilot projects', http://www.sdpc.gov.cn/zcfb/zcfbtz/2010tz/t20100810_365264.htm (in Chinese)

[6]See http://www.ndrc.gov.cn/gzdt/t20121205_517506.htm (in Chinese)

[7]See 'Greening up Qingdao', *China Daily*, 12 January 2009, http://www.chinadaily.com.cn/bw/2009-01/12/content_7386366.htm

Table 1 Low carbon measures taken by Chinese cities: some selected measures

Region	Cities	Illustrative measures
Bohai Rim region	Beijing	The city is to promote greening systems in production, consumption and environment protection, by implementing nine programmes in areas including energy, building, transport, air, solid waste, water and ecology
	Dezhou	Based on its industrial advantages as the 'China Solar City', the city is to accelerate application of solar energy in various areas
	Baoding	The city is focused on development of industrial parks and industrial clusters focusing on wind power, PV solar, energy saving, energy storage, electric power transmission and transformation, and electric automation equipment manufacturing etc.
Yangtze River Delta region	Shanghai	By applying the low carbon concept that has been implemented in the Shanghai 2010 Expo, the city plans to complete three low carbon demonstration areas in Songming, Lingang and Hongqian Business District, and a Dongtan Eco-city project during the period of 12th Five Year (2011–2015)
	Nanjing	The city is to increase the proportion of renewable energy in total energy consumption by encouraging 'green consumption'
	Hangzhou	The city has established a comprehensive development goal toward low carbon economy, low carbon buildings, low carbon transport, low carbon life style, low carbon environment and low carbon society
Pearl River Delta region	Shenzhen	The city is to focus on energy saving in key areas, to use public funding for promoting investment in low carbon development, and to develop three strategic industries including biology, new energy and Internet-related industries
	Zhuhai	The city is to identify key areas in carbon emission reduction and establish solid reduction targets
	Nanchang	The city is to build a number of low carbon demonstration industrial parks. The city is also to provide its residents free bicycles to encourage low carbon transport
Southeast China region	Chongqing	The city is to develop low carbon economy in adjunct with its industrial structure change, urban planning and technological innovation, and to increase the share of energy-saving and environmental protection industries in the economy
	Chengdu	The city is to build a low-carbon economic development experimental zone, zero-carbon agricultural demonstration zone, and zero-carbon tourism demonstration zone. The city is also to establish and improve its public ecological compensation mechanism, and compensation and incentive mechanisms based on energy consumption per unit GDP
	Guiyang	The city has established the first court in the country specialized in legal cases in relation to environmental protection; and has issued the first local regulation on ecologic progress. The city has also converted all its buses from petrol engines to LPG engines

Source: Based on Chen et al. [7] and Tian [8]

these local-level initiatives. Under the 12th FYP (2011–2015) and the current 13th FYP (2016–2020), there are chapters on CE initiatives and further development; the State Council (equivalent to the Cabinet) issued a series of proposals to implement the Circular Economy goals, followed up by a series of more detailed regulations issued by the ND&RC. This is a typical sequence found in China—a general goal or aspiration (as in the 12th FYP and the current 13th FYP) is followed up with a decision by the State Council to embark on serious implementation, with guidelines and regulations then being issued by the ND&RC. This sequence ensures that China's greening of its cities is being pursued in a systematic and serious way with legislative backing for administrative and financial promotion. A particularly good example of this process is found near the port city of Tianjin.

4 Tianjin

The Sino-Singapore Tianjin Eco-City (SSTEC) is a new eco-city with sustainable features built in from the start, involving a collaboration between the Chinese and Singapore governments. The SSTEC is situated within the Tianjin-Binhai New Area, a fast-growing industrial region located in the Bohai Bay area and now identified as a third industrial engine in China behind the Pearl River delta (featuring the cities Guangdong and Shenzhen) and the Yangtze River Delta (featuring Shanghai and Suzhou). Its rate of industrial growth is more than twice the national average. Within this cluster of industries the SSTEC is designed as a fresh start for a city drawing on the experience in recycling and resource efficiency already developed by Singapore. The eco-city, designed to have a population of 350,000, is located 40 km from Tianjin city centre, which is in turn located just 110 km from Beijing and connected by a very fast high-speed rail service, the first and still most significant in China.

Following earlier collaboration between Singapore and China over the Suzhou technology park, a new agreement was reached between the governments in 2007 for the creation of an eco-city near Tianjin. The ground-breaking ceremony was held in September 2008. This has been followed by new memoranda of agreement and the adoption of standards consistent with Singapore's own standards for water, waste and energy renewal. The Singapore government has formed an Inter-Ministerial Committee to coordinate its input, as a sign of the importance attached by Singapore to the eco-city's success. The Master Plan for the city has been prepared by the Singapore Urban Redevelopment Authority (URA). As a planned eco-city, SSTEC has a number of metrics that ensure its development proceeds along sustainable lines. Amongst these are: close-to-zero carbon emissions; all buildings to qualify as 'green' in terms of renewable energy and water recycling; overall solid waste recycling to reach 60 %; and 100 % water recycling. Particular features are the installation of an underground vacuum-driven waste disposal system for the city (a world first) and experiments with driverless automatically guided vehicles (supplied by Google). Charging stations for electric vehicles are to be installed at every major intersection.[8]

One of us (MCH) visited the SSTEC in June 2011 and September 2012 having the chance to walk around the city. It had been built on an unpromising reserve of industrial wasteland, on the principle that if an eco-city could be built *here*, then it could be built anywhere. The city is now half-built and still has a drab feel to it—but with noticeable drive and ambition. Walking down the central avenue it is easy to see how the different facets of green development (lighting, water recycling, waste disposal) have all been integrated, while the buildings are ultra-clean and all are equipped with solar panels. It is a city with grand ambitions. A key aspect of the city's development involves the smart grid, providing both an experimental city-wide implementation zone as well as an opportunity for large corporates to test their latest technologies and designs. The Chinese white goods manufacturer Haier, for example, is developing new standards for the operation of the green home or smart house Energy Management System (EMS) as part of the architecture of the smart grid—promising it leadership of this emerging huge market both in China and internationally [11]. This is what China calls its 'indigenous' innovation system, promising to drive the country's efforts to shift from imitation to innovation.

The SSTEC has attracted much international attention, including a World Bank study in its early phases [12]. Particular attention is being paid to the capacity of the eco-city to attract lower-cost financing (because of its green credentials) and its capacity to capture latecomer (and first mover) advantages in having entire systems that are based on renewable energies, water and waste recycling [13]. Ultimately the goal of China's green city strategy is to demonstrate that eco-cities carry a cost and competitive advantage over their older, smoke-filled industrial predecessors.

5 China's Greening City Strategy

China clearly recognizes that its urbanization and industrialization are twin revolutions, both setting unprecedented challenges as well as opportunities to build an alternative, green industrial and urban model. To combat and complement the conventional industrialization model, involving fossil fuels and extensive resource throughput, China is seen to be making huge efforts to build a green alternative—starting with energy and water renewal and encompassing not just power plants and factories but whole cities and regions in the measures being taken. International

[8]See Coco Liu, 'China's city of the future rises on a wasteland', *New York Times*, September 28, 2011, http://www.nytimes.com/cwire/2011/09/28/28climatewire-chinas-city-of-the-future-rises-on-a-wastela-76934.html?pagewanted=all

agencies and organizations as well as international consultancies are keeping close tabs on China's efforts to green its cities.[9]

The McKinsey 2009 report [11] on China's urbanization identifies four possible models for China's greening of cities, namely (1) a small number of super-cities (such as Beijing and Shanghai); (2) a hub-and-spoke model, involving two or three hubs and several smaller cities clustering around them; (3) distributed growth involving a large number of medium-sized cities; and (4) unplanned urbanization involving lots of smaller towns competing with each other. Taking this as a convenient framework, we see that the SSTEC fits within a hub-and-spoke model with Tianjin as one of the hubs and smaller entities clustering around it, with SSTEC providing the template for a fresh start. The model encourages resource-sharing in the way of all clusters and networks, but applied at a larger scale. The idea is that core centres of green development will be initiated and then expand and make connections with each other—rather like the stones in the ancient Chinese board game GO, where strength is found through connection rather than through stand-alone policies.

The outcome of China's efforts to green its cities, and thereby green its twenty-first century economy, is one of the great 'uncontrolled' experiments of the twenty-first century—uncontrolled in a social scientific sense, of conducting a process without a 'control group' for comparison. The outcome is anything but determined. But there are grounds for cautious optimism in the fact that China's development is increasingly urbanized, with the energy and efficiency gains that can be captured by smart catch-up strategies.

6 Chapter Summary

- China is embarked on a major industrialization and urbanization programme, equivalent to building seven new cities of three million inhabitants each, every year.
- The environmental costs of following a 'Business as usual' model would be prohibitive, and so China is experimenting with an alternative 'low-carbon cities' model.
- The 'low-carbon cities' programme is driven by local initiatives but coordinated as a central planning goal by the National development and Reform Commission.
- China is capturing latecomer advantages by leapfrogging to advanced low-carbon city designs, following a greening model of development that maximizes the diffusion of new approaches to urban design.

[9] See OECD 2013 report 'Urbanisation and green growth in China' which gives a comprehensive update [14], and World Bank reports including *Eco*^{2} *cities: Ecological cities as economic cities* (WB 2010) [15], and *Building sustainability in an urbanizing world* (WB 2013) [16]. McKinsey has been a leading international consultancy examining the greening of China's cities, as in [11, 17, 18].

- There are grounds for cautious optimism in that industrial development that is increasingly urbanized promises energy and resource efficiencies that can be captured by smart catch-up strategies.

References

1. NBS: Chinese Statistics Yearbook. National Bureau of Statistics, Beijing, China (2015)
2. Mathews, J.A., Tan, H.: China's Renewable Energy Revolution. Palgrave Pivot, London (2015)
3. Wei, Y.M., Liu, L.C., Fan, Y., Wu, G.: The impact of lifestyle on energy use and CO^2 emission: An empirical analysis of China's residents. Energy Policy. **36**(1), 247–257 (2007)
4. Su, M.R., Chen, B., Xing, T., Chen, C., Yang, Z.F.: Development of low-carbon city in China: Where will it go? Procedia Environ. Sci. **13**, 1143–1148 (2012)
5. Zhou, N., He, G., Williams, C.: China's development of low-carbon eco-cities and associated indicator systems. In: Paper LBNL-5873E. Lawrence Berkeley National Laboratory, Berkeley, CA (2014)
6. Yu, L.: Low carbon eco-city: New approach for Chinese urbanisation. Habitat Int. **44**, 102–110 (2014)
7. Chen, C., Liu, W., Liu, L., Huan, H.: A study on the development strategy of Chinese low carbon cities. Ecol. Econ. **7**(1), 10–15 (2013) (in Chinese)
8. Tian, M.: Guiyang: From the 'acid rain city' to a low-carbon city. Economic Daily. (in Chinese) (2012)
9. Guo, W.: Qingdao resolves "five residue" problems left by history through developing the industry chain of circular economy. Renew. Resour. Recycling Econ. **3**, 30–41 (2009) (in Chinese)
10. IEK: Urbanization under the 12th Five-Year plan. Industrial Technology Research Institute, Hsinchu, Taiwan (in Chinese) (2011)
11. McKinsey: China's green revolution: Prioritizing technologies to achieve energy and environmental sustainability. www.mckinsey.com/~/.../sustainability/.../china_green_revolution.ashx (2009)
12. World Bank: Sino-Singapore Tianjin Eco-City: A case study of an emerging eco-city in China. http://www-wds.worldbank.org/external/default/WDSContentServer/WDSP/IB/2011/01/17/000333037_20110117011432/Rendered/PDF/590120WP0P114811REPORT0FINAL1EN1WEB.pdf (2009)
13. Hu, M.C., CY, W., Shih, T.: Creating a new socio-technical regime in China: evidence from the Sino-Singapore Tianjin Eco-City. Futures. **70**, 1–12 (2015)
14. OECD: Urbanisation and green growth in China. In: OECD Regional Development Working Papers, 2013/07. OECD, Paris (2013)
15. World Bank: Eco^2 cities: Ecological cities as economic cities. http://web.worldbank.org/WBSITE/EXTERNAL/TOPICS/EXTURBANDEVELOPMENT/0,,contentMDK:22643153~pagePK:148956~piPK:216618~theSitePK:337178,00.html (2010)
16. World Bank: Building sustainability in an urbanizing world. http://www.worldbank.org/en/topic/urbandevelopment/publication/building-sustainability-urbanizing-world (2013)
17. McKinsey: China's next chapter, McKinsey Quarterly. http://www.mckinsey.com/insights/asia-pacific/chinas_next_chapter (2013)
18. Woetzel, J.: How green are China's cities? McKinsey Quarterly. http://www.mckinsey.com/insights/energy_resources_materials/how_green_are_chinas_cities (2011)

Low-Carbon Urban Infrastructure

Stephanie B. Ohshita and Nan Zhou

Abstract As cities consider how to provide essential urban services in the post-fossil fuel age, they find that incremental strategies are not enough. The challenge, and opportunity, is to reinvent essential city infrastructure—for water, food, shelter, energy, transport, culture, and economy—in a climate-friendly way. By reclaiming human-scale neighborhoods, relearning and further developing sustainable and passive building techniques, and reaching forward with technology in the service of society, cities can thrive with less energy and fewer greenhouse gas (GHG) emissions. This chapter examines essential concepts and examples of low-carbon urban infrastructure, highlighting urban form designed in harmony with the city's geography; resilience to climate change impacts as well as reductions in greenhouse gas emissions; and prioritization of demand-side management in city systems, through improved design, efficiency, and de-carbonization.

Keywords Low-carbon cities • Urban infrastructure • Urban form • Urban resilience • Demand-side management

Key Terms

Low-Carbon Urban Infrastructure: the physical facilities supporting the functioning of a city—such as transportation systems, energy systems, building systems, water, and wastewater systems—that are designed to use less energy and emit fewer greenhouse gasses.

Urban Form: the spatial imprint of a city, including natural topography, design of public spaces and built environment, development density, and transportation networks.

Resilience: the ability of a system or community to manage climate change impacts and adapt to climate variability and extremes.

Demand-Side Management (DSM): management of energy consumption patterns to achieve large-scale energy savings (efficiency and conservation) and reduce peak

S.B. Ohshita (✉) • N. Zhou
Lawrence Berkeley National Laboratory, Berkeley, CA, USA
e-mail: SBOhshita@lbl.gov; NZhou@lbl.gov

S. Dhakal, M. Ruth (eds.), *Creating Low Carbon Cities*,
DOI 10.1007/978-3-319-49730-3_10

electricity demand (load shifting), through use of improved technologies and utility and government programs.

1 Urban Form: Foundation of Low-Carbon Infrastructure

The demands on the infrastructure of a city, and in turn its carbon footprint, are influenced by the size of the population, the socioeconomic mix, and the overall design of a city—*urban form* [1, 2]. One favorable urban form for low-carbon cities is that of mixed-use zoning, combining residential and commercial and public uses in clusters [3], or *urban villages*, which are a type of urban neighborhood that emphasizes human-scale, mixed-use design [4, 5]. From Portland's "20-minute neighborhood" [6] to the UK's "Proximity Principle" [7], residents of urban villages are able to access daily needs of housing and public space, food markets and restaurants, shops and service businesses, and schools and parks, by walking or bicycling. Workplaces and other activities can be accessed via well-designed public transit corridors. Moderately high-density urban villages have less demand for motorized transport, and interconnected urban villages facilitate public transport infrastructure, which lessens citywide energy consumption and greenhouse gas emissions [5].

In contrast, a sprawling urban form with long-distance vehicle commuting can result in as much as four times more vehicle miles travelled (VMT) compared to high-density urban areas, and three times higher household carbon dioxide equivalent (tCO_2e) emissions overall: 21 tCO_2e/household compared to 6 tCO_2e/household in US cities [1]. For existing urban neighborhoods that shift to mixed-use zoning and other low-carbon transport strategies, cities may achieve 30 % savings in VMT and CO_2e within 10–20 years [6].

For developing cities where the population is growing rapidly, urban form and infrastructure choices have a significant and long-lasting impact on urban energy consumption and GHG emissions. Indian cities of less than one million population typically have a high-density [>150 persons per hectare], mixed-use urban form where a high share of trips is accomplished by walking or bicycling [8]. As these cities grow, trip lengths can become longer, thereby increasing the demand for motorized transport—and energy and carbon. By clustering development to maintain densities greater than 50 persons per hectare, and investing in public transit infrastructure, development can take a low-carbon route, as achieved in cities in Japan, Europe, and elsewhere [8]. However, if investment is channeled into automobile-focused road infrastructure, the urban form of the city takes an unfavorable high-carbon turn, as evidenced in many US cities.

The land development patterns of a city also determine the location and types of residential and commercial buildings that are developed, which in turn influence energy consumption. Orienting buildings along east-west streets, with the most occupied portions of buildings optimized for solar gain or shading, makes direct use of the ultimate energy source: the sun. Solar orientation of buildings has been practiced since antiquity and was formalized by the ancient Chinese and Greek civilizations [9]. Multiunit dwellings typically have lower per capita energy use than single-family houses. Analysis in Toronto neighborhoods found a tenfold variation, with multifamily units near public transit and services emitting 1.3 tCO_2e per capita, while single-family homes in distant, automobile-dependent, sprawled developments emitting 13.0 tCO_2e per capita [10].

With human disruption of the Earth's climate system, city design cannot rely on past patterns of climate, but must look ahead and consider future climate change scenarios in its infrastructure planning [11]. *Resilience* [12] becomes a more important design criteria in a destabilized climate. The City of Chicago has already begun planting tree species based on future climate scenarios, to maintain urban appeal, to provide greater summer cooling for people and buildings and streets, and to prevent erosion and slow storm-water run-off [13]. Coastal cities are conducting climate change vulnerability assessments for urban infrastructure [14]. Even as cities pursue highly efficient buildings, and biodiesel and electric bus systems, they cannot achieve low-carbon benefits if infrastructure is developed, for example, in a floodplain likely to be inundated by sea-level rise and storm surges.

2 Energy Infrastructure: Demand and Supply

For low-carbon energy infrastructure, we begin with the end: energy end use, *demand-side management* [15]. For what purposes are we using energy—i.e., what energy services does a city need—and how can energy infrastructure support social goals? By reexamining ways to provide energy services through improved development patterns, and better design of buildings and other urban infrastructure, cities can resolve the conundrum of balancing energy demand with low-carbon energy supply.

Thus the first priority in urban energy infrastructure is energy conservation through urban design (Table 1). There is no need to increase the supply of heating oil or natural gas if buildings are designed with passive solar heating and well-insulated thermal envelopes. And the energy savings are achieved every year, unlike supply-side approaches.

End-Use Efficiency

The next strategy for a low-carbon urban energy system is energy-efficient buildings and appliances, from heating and cooling to washing and computing. Though not energy infrastructure in the usual sense, energy-efficient buildings have a direct influence on the energy supply infrastructure needed for a city [16]. Efficiency standards for municipal operations, combined with standards for building types

Table 1 Low-carbon energy infrastructure strategies

Strategy	Highlights
Energy Conservation in Urban Development and Building Design	Low-demand and Zero Net Energy systems, including passive energy techniques such as building orientation, solar gain, thermal mass, shading, daylighting, passive ventilation
Energy Efficiency in End-Use Systems	Energy-efficient building systems and appliances: building envelope, space heating and cooling, ventilation, water heating and cooling, lighting, cooking and refrigeration, other electric appliances
Supply-side Efficiency in Energy Conversion	Heat recovery in thermal processes for electricity generation (cogeneration), and utilization in municipal heating. Improved efficiencies in municipal-scale electricity, in combined heat and power (CHP) or renewable generation
Distributed Energy Systems	Localized electric power generation technologies, combined with load management, energy storage, and local dispatch systems. Often utilized to increase renewable power in energy supply, avoid transmissions losses, and provide greater power reliability
Renewable Energy Infrastructure	To complement investments in renewable power generation, investment in grid and dispatch technologies is important for increasing the share of renewable power and balancing variability in electricity generation from renewable sources

Source: Authors

across the city, can yield large energy and carbon savings [17]. Energy conservation and efficiency can also help cities manage tight budgets. Investment in municipal building retrofits in Los Angeles was recouped in 3 years from energy savings [10]. Energy efficiency improvements in street lighting, by replacing 140,000 fixtures with light-emitting diode (LED) lights, are saving the city US$ 10 million and 40,500 tCO_2e per year [10]. These ongoing savings in carbon and energy bills free up financial resources for other municipal needs.

Supply-Side Efficiency and De-carbonization

On the supply side, where electricity and heat are preferred forms of urban energy, improvements in conversion efficiency can yield more energy from existing supply. Investment in cogeneration of electricity and heat (combined heat and power, CHP) generates substantial savings of input energy and related GHG emissions. Whereas most fossil-fired electricity generation has a conversion efficiency near 30 %, CHP, by utilizing waste heat, typically converts 75–80 % of input energy into useful electricity and heat [18]. To more swiftly improve supply efficiencies and integrate more renewable electricity generation, municipalities and the private and commercial facilities within them are turning to distributed energy resources (DER), including distributed generation (DG) [19]. These smaller scale energy resources can connect directly to the local grid, avoiding transmission losses and improving reliability. To further support the goal of de-carbonization in energy supply, municipal utilities in Germany and elsewhere are pushing ahead toward 100 % renewable electricity goals. Munich's local utility, Stadtwerke München (SWM), is on track to supply its 1.4 million residents the electric subway and light rail, and eventually industrial customers with all renewable electricity by 2025 [20].

3 Water Infrastructure: Low-Carbon and Resilient

Urban water infrastructure is energy consuming and GHG emitting, in the conveyance, heating, and treatment of water supply, wastewater, and storm water. The state of California estimates that 19 % of electricity, 32 % of natural gas, and 88 million gallons of diesel fuel are consumed annually in connection with water use, with a major share of energy and emissions due to urban water use [21].

At the same time, water infrastructure is being strongly impacted by changes in the climate system [22]. Declining snow packs affect the timing and volume of freshwater flows, and intense storms impact water quality and overwhelm conveyance and treatment systems. Rising sea levels and hotter temperatures lead to increased salinity in water supplies. Fluctuating precipitation cycles reduce groundwater recharge, further impacting limited supplies. Thus water infrastructure must endeavor to be both low carbon and resilient. Table 2 highlights strategies for low-carbon and resilient urban water infrastructure. Two of these strategies are discussed further below.

Table 2 Low-carbon and resilient strategies for urban water infrastructure

Save first: end-use water conservation and efficiency	Lessen end-use demand for water through more efficient appliances (washer, showers, toilets); improved commercial and industrial processes; less wasteful watering techniques for green spaces and agriculture. Establish water consumption limits before supplies are (further) impacted
Low-energy water supply	Reduce losses in water supply systems from evaporation and leaks in conveyance channels and piping. Choose low energy-intensity options for any new or modified supply while minimizing other environmental impacts of water extraction
Water reclamation	Implement grey-water systems, industrial water recycling, and rainwater collection for greater utilization of limited water supply
Wastewater treatment and biogas cogeneration	Capture methane from wastewater treatment to avoid emissions and to utilize for cogeneration of electricity and heat or production of biogas
Resilience in water infrastructure	Prepare for projected changes in water cycles and other climate impacts in all water infrastructure: green roofs, bio-swales, and permeable pavement to manage storm water; enhanced treatment to manage changing water quality; diversity in water supply; water management and rationing plans for droughts, floods, and other emergencies

Save First

As with urban energy, low-carbon water infrastructure would do well to prioritize demand-side saving of water, coupled with efficient and renewable energy use related to water. This prioritization of strategies reduces the demand for water and energy, typically at a lower cost than pursuing increased supplies, which may not even be available. As an example, consider the demand for residential hot water, for washing people, clothing, and other textiles [23]. Use of low-flow showerheads and water-efficient washing machines conserves hot water, as does taking shorter showers and washing textiles with cold water. The water conservation efforts can then be coupled with low-carbon and efficient energy strategies: solar thermal water heating, supplemented with on-demand electric or gas-fired water heating as needed to bring water temperature up to a desired level [21]. When this combination of strategies is implemented citywide, the savings in water, energy, and carbon can be significant [21].

Wastewater Treatment and Biogas Cogeneration

On the other end of the urban water cycle is wastewater treatment. From a GHG emission perspective, methane released from the decomposition of organic matter in sewage is as problematic as emissions from energy consumption during treatment. Recognizing the opportunity to both prevent GHG emissions and recover a valuable energy source, municipal wastewater utilities from the Philippines to Philadelphia are investing in infrastructure to capture and utilize methane-containing biogas from anaerobic digestion. In the USA roughly 25 % of wastewater treatment utilities generate electricity from digester biogas, often with cogeneration (combined heat and power, CHP) facilities [24]. A smaller share of utilities cleans the gas to produce bio-methane. The combination of digester biogas and CHP has multiple benefits: low-cost electricity production, reduced fuel purchases for electricity and heating, reduced emissions of GHGs and other pollutants, and possible recognition as a renewable energy source under state and national policies [24].

4 Transportation Infrastructure: Rethinking Mobility

Globally, greenhouse gas (GHG) emissions from the transportation sector account for 22 % of the total and are growing, with road transport—cars and light trucks—being the dominant mode of transport [25]. A systems view of transportation and carbon tells us that we must rethink mobility, since current patterns of passenger and freight transport are at odds with de-carbonization [26]. Where are we going

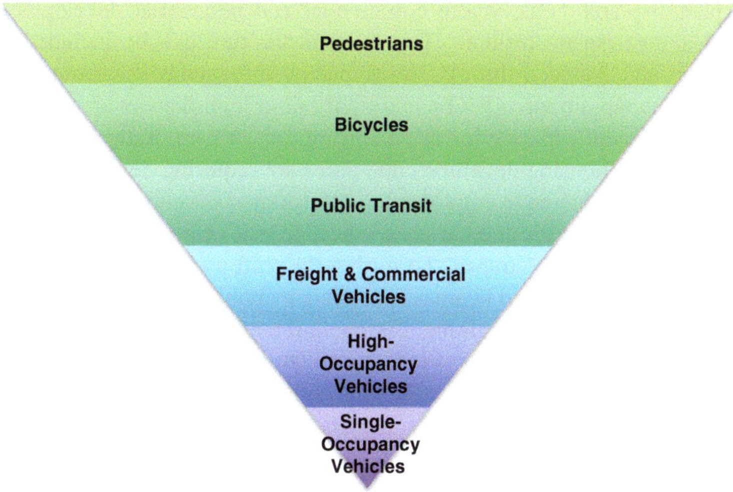

Fig. 1 Priorities for low-carbon transport modes. Source: Based on hierarchy of transport modes in [6].

and why? How can we refashion mobility and access for communities, especially in the world's growing cities? How can we reconsider transportation infrastructure that prioritizes people and a rapidly changing natural environment, beyond the age of the fossil-fired internal combustion engine?

Prioritize Investments in Low-Carbon Transport Modes

A first step in low-carbon transport is literally that—a step on foot—with the prioritization of infrastructure for walking and bicycling [27]. From New York and Portland to Buenos Aires and Guangzhou, cities around the world are utilizing a hierarchy of transportation modes to reduce energy consumption and GHG emissions in the transportation sector (ITDP [17, 28]). Figure 1 shows transportation modes in order of low-carbon priority [6].

Transit-Oriented Development (TOD)

From Tokyo to Curitiba, many of the world's cities have found that transit-oriented development (TOD) is a cleaner, more efficient, and more accessible transportation strategy than plans oriented toward private vehicles [17]. Vehicle miles (or kilometers) travelled (VMT) and GHG emissions can be reduced with integrated transit planning and TOD, where commercial and residential developments are clustered along transit corridors. Transit infrastructure can be included in new construction and financed through development agreements. Investment in transit—buses, bus rapid transit (BRT), and light rail in high-density areas—is needed to provide sufficient capacity for those shifting out of cars. Rather than requiring parking spaces in housing developments, cities are requiring developers to bundle transit passes and encourage employers to offer transit benefits rather than parking [29]. Parking for bicycles and car shares near transit centers enables residents to better connect with public transit. To further the shift away from private vehicles, cities are raising parking fees in areas accessible by public transit and implementing programs to park and ride. Walking, biking, and public transit are more appealing with easy access and payment systems, and with well-crafted transit information and public outreach. Establishment of city targets for a desired mix of transport modes, along with public progress reports, focuses attention on the shift to low-carbon transport modes [6].

Integrated Transport Infrastructure

These low-carbon transport modes must be integrated, as well as prioritized, in infrastructure investment. Integrated transport planning for low-carbon development has the goal of enhancing a community's accessibility to resources and services with (1) low-VMT transport, i.e., urban development and transport options that reduce the vehicle miles travelled per person and in total, and (2) low-carbon transport modes, from non-motorized transport to efficient, clean-powered vehicles. Low-carbon transportation infrastructure in Guangzhou highlights the importance of integrating public transit with walking and biking. The Guangzhou BRT is the first BRT in Asia connected with the metro rail system [28, 30]. The Guangzhou BRT system includes bicycle parking in its station design and a greenway parallel to the corridor, integrating the city's bike share program of nearly 5000 bicycles and 50 bike stations ([30]; National [31]). The BRT system carries more passengers per hour than any mainland Chinese metro outside of Beijing, tripling the capacity reached by other BRT in Asia [30]. The efficiency improvements from BRT have reduced travel time for bus riders and motorists along the route by 29 % and 20 %, respectively. The fuel savings will in turn amount to 86,000 tCO_2e annually [30]. As another example, the bike share system in Hangzhou—China's first bike share and one of the largest in the world—is designed to support non-motorized transport and to connect travelers with public transit [32].

Multimodal Streets

Multimodal Streets, also known as "Complete Streets," exemplify integrated transport infrastructure for low-carbon urban form and mobility. Complete Streets aim to balance multiple transport modes and create appealing urban spaces [33]. Design features include lane striping and signage, raised crosswalks and pedestrian control signals, bus pullouts, and traffic calming measures [34]. Complete Streets have sidewalks that easily access retail, restaurants, and other pedestrian services—not forbidding concrete building fronts along barren superblocks [35]. Trees and vegetation, shaded entrance ways, umbrellas, and benches all contribute to pedestrian appeal and safety, along with reduced transport pollution [36].

New York City started work on Complete Streets with small pilot projects to make streets more humane, utilizing paint and chairs to create more pedestrian-friendly spaces adjacent to streets, and making changes in crossing signals to more safely integrate walking and driving modes [36]. The city then scaled up to larger infrastructure projects, converting automobile lanes to bus rapid transit lanes and physically separated bicycle lanes. As a result, traffic congestion was reduced on the improved multimodal streets by 10–50 %, bus speed and ridership improved by 10–50 %, and local businesses saw sales increase by as much as 50 % [37].

Cleaner Vehicle Technology

Attention is often given to new technological approaches for the transportation sector, such as improved fuel economy in conventional vehicles, hybrid fuel electric vehicles, fully electric vehicles (EV), or hydrogen fuel cell vehicles, because of the potentially large savings in operational GHG emissions per vehicle. Certainly, cities can achieve carbon savings with investments in vehicles with improved fuel economy, or reduced GHG per vehicle mile. The new European Union vehicle CO_2 emission standard will bring down emissions from new cars to 130 g CO_2/km by 2015, and down to 95 g CO_2/km by 2020 [38]. Hybrid vehicles emit roughly half the GHG emissions of a typical passenger car [29]. For EV and hydrogen vehicles, the savings are contingent on the source of electricity or hydrogen. Coal-fired electricity and natural gas-derived hydrogen are not low-carbon energy sources, whereas an EV powered by renewable electricity can save as much as 70 % GHG emissions compared to a typical gasoline-fired car [29].

From a systems perspective, however, new vehicle technology does not go far enough in curbing GHG emissions, as it propagates a car-focused transportation system [26]. The operational energy utilized per person on bus or light rail can be 80 % lower than one person driving a car alone [29]. The energy expended per person by walking or biking is an even smaller fraction of the energy required for a single-occupancy vehicle. And when infrastructure and embodied energy are considered, walking, biking, and public transit are even more appealing for low-carbon cities [39]. Walking, biking, and public transit infrastructure also contribute multiple social and environmental benefits, including greater access to mobility, greater social interaction, enhanced economic productivity, and improved health through active transport and better air quality [40, 41]. For all these reasons, prioritizing investment in the lowest carbon transport modes is needed for transportation infrastructure.

5 Conclusion

- With increasing variability and extremes due to disruption of the climate system, low-carbon infrastructure must be resilient as well as minimize greenhouse gas emissions.
- Essential criteria for low-carbon urban infrastructure include an urban form of mixed-use zoning and interconnected urban villages, along with attention to clustered population density and the socioeconomic patterns of the city.
- Urban energy infrastructure must prioritize efficient use of energy (demand-side management), even as investments are made in renewable and distributed energy systems (supply-side de-carbonization).
- The strong connection among water use, energy, and greenhouse gas emissions necessitates efficient water infrastructure throughout the cycle of urban water use, especially in water heating and wastewater treatment.
- To curb the upward trend in transportation energy and GHG emissions, cities must give greater attention to social needs for mobility, shift away from car-focused transport, and prioritize non-motorized and public transit infrastructure.

Acknowledgements Some of the research for this chapter was supported by the China Sustainable Energy Program of the Energy Foundation through the US Department of Energy under Contract No. DE-AC02-05CH11231. Thanks to the University of San Francisco Faculty Development Fund and student Sarah Lund for research assistance.

Disclosure This work was supported by the China Sustainable Energy Program of the Energy Foundation through the US Department of Energy under Contract No. DE-AC02-05CH11231.

The authors have no other relevant affiliations or financial involvement with any organization or entity with a financial interest in or financial conflict with the subject matter or materials discussed in the manuscript apart from those disclosed. No writing assistance was utilized in the production of this manuscript.

References

1. Calthorpe, P.: Urbanism in the Age of Climate Change. Island Press, New York (2010)
2. Echenique, M.H., Hargreaves, A.J., Mitchell, G., Namdeo, A.: Growing cities sustainably J. Am. Plann. Assoc. **78**(2), 121–137 (2012)
3. Suzuki, H., Cervero, R., Iuchi, K.: Transforming Cities with Transit: Transit and Land-Use Integration for Sustainable Urban Development. World Bank, Washington, DC (2013)

4. Franklin, B., Tait, M.: Constructing an image: the urban village concept in the UK. Plann. Theory. **1**(3), 250–272 (2002)
5. Paelke, R.: Sustainable development and urban life in North America. In: Vig, N., Kraft, M. (eds.) Chapter 11 in Environmental Policy, 7th edn. CQ Press, Washington, DC (2010)
6. City of Portland and Multinomah County: Climate action plan. Portland, OR, USA. https://www.portlandoregon.gov/bps/49989 (2009)
7. Buchan, K.: A low carbon transport policy for the UK. http://www.bettertransport.org.uk/media/press_releases/november_2008/cutting_emissions (2008)
8. Dhar, S., Pathak, M., Shukla, P.R.: Low carbon city: a guidebook for city planners and practitioners. Report for the UNEP project promoting low carbon transport in India. (2013)
9. Butti, K., Perlin, J.: A Golden Thread: 2500 Years of Solar Architecture. Cheshire Books, Palo Alto (1980)
10. World Bank: Cities and climate change: an urgent agenda. http://siteresources.worldbank.org/INTUWM/Resources/340232-1205330656272/CitiesandClimateChange.pdf (2010)
11. McGregor, A., Roberts, C., Cousins, F.: Two Degrees: The Built Environment and Our Changing Climate. Routledge, New York (2013)
12. Levina, E., Tirpak, D.: Adaptation to Climate Change: Key Terms. Paris, OECD and IEA (2006)
13. City of Chicago: Chicago climate action plan. Chicago, IL, USA. www.chicagoclimateaction.org (2009)
14. San Francisco Bay Conservation and Development Commission (SF BCDC): Assessing climate change vulnerability and risk. Report of the ART program, Adapting to Rising Tides. San Francisco Bay Conservation and Development Commission, San Francisco. http://www.adaptingtorisingtides.org/project-reports/ (2011)
15. Davito, B., Tai, H., Uhlenr, R.: The Smart Grid and the Promise of Demand-Side Management. McKinsey (2010)
16. Seto, K.C., Dhakal, S., Bigio, A., Blanco, H., Delgado, G.C., Dewar, D., Huang, L., Inaba, A., Kansal, A., Lwasa, S., McMahon, J.E., Müller, D.B., Murakami, J., Nagendra, H., Ramaswami, A.: Human settlements, infrastructure and spatial planning. In: Edenhofer, O., Pichs-Madruga, R., Sokona, Y., Farahani, E., Kadner, S., Seyboth, K., Adler, A., Baum, I., Brunner, S., Eickemeier, P., Kriemann, B., Savolainen, J., Schlömer, S., von Stechow, C., Zwickel, T., Minx, J.C. (eds.) Climate Change 2014: Mitigation of Climate Change. Contribution of Working Group III to the Fifth Assessment Report of the Intergovernmental Panel on Climate Change. Cambridge University Press, Cambridge (2014)
17. Zhou, N., Price, L., Fridley, D., et al.: Strategies for local low-carbon development. Report. Lawrence Berkeley National Laboratory, Berkeley, CA. http://china.lbl.gov/publications/strategies-local-low-carbon-developme (2012)
18. International Energy Agency (IEA): Cogeneration and District Energy. Report prepared by Tom Kerr. IEA/OECD, Paris (2009)
19. Akorede, M.F., Hizam, H., Pouresmail, E.: Distributed energy resources and benefits to the environment. Renew. Sustain. Energy Rev. **14**, 724–734 (2010)
20. Yapp, R.: City utilities push Germany's switch to renewables. Renewable Energy World. 4 October. http://www.renewableenergyworld.com/rea/news/article/2012/10/city-utilities-push-germanys-switch-to-renewables (2012)
21. California Energy Commission (CEC): California's Water-Energy Relationship. Prepared in Support of the 2005 Integrated Energy Policy Report Proceeding. Report CEC-700-2005-011-SF (2005)
22. Asian Development Bank (ADB): Increasing climate change resilience of urban water infrastructure. Guidebook. Asian Development Bank, Manila. http://www.adb.org/publications/guidebook-increasing-climate-change-resilience-urban-water-infrastructure (2013)
23. Siddiqi, A., de Weck, O.: Quantifying end-use energy intensity of the urban water cycle. J. Infrastruct. Syst. **19**(4), 474–485 (2013)
24. U.S. Environmental Protection Agency (U.S. EPA): CHP Partnership. Opportunities for combined heat and power at wastewater treatment facilities: market analysis and lessons from the field. Report prepared by: Eastern Research Group, Inc. (ERG) and Resource

Dynamics Corporation (RDC) for the U.S. Environmental Protection Agency, Combined Heat and Power Partnership. October (2011)
25. International Energy Agency (IEA): CO_2 from Fossil Fuel Combustion. IEA, Paris (2011)
26. Tyfield, D.: Transportation and low carbon development. In: Urban, F., Nordensvard, J. (eds.) Chapter 13 in Low Carbon Development: Key Issues. Earthscan from Routledge, London (2013)
27. Chicago Department of Transportation (Chicago DOT): Complete streets Chicago: design guidelines. http://chicagocompletestreets.org/ (2013)
28. Institute for Transportation Development and Policy (ITDP): Sustainable transport award. ITDP. http://www.itdp.org/get-involved/sustainable-transport-award/ (2011)
29. San Francisco Metropolitan Transportation Agency (SF MTA): Climate action strategy. San Francisco Metropolitan Transportation Agency (2011)
30. Hughes, C., Zhu, X.: Guangzhou, China, Bus rapid transit: emissions impact analysis. Institute for Transportation and Development Policy (ITDP). May. http://www.itdp.org/news/guangzhou-brt-reduces-co2-emissions-by-86000-tonnes-annually (2011)
31. National Geographic: Guangzhou, China, wins sustainable transport prize. National Geographic Daily News. 24 January. http://news.nationalgeographic.com/news/energy/2011/01/110124-guangzhou-china-sustainable-transport-prize/ (2011)
32. Shaheen, S., Zhang, H., Martin, E., Guzman, S.: Hangzhou public bicycle: understanding early adoption and behavioral response to bikesharing in Hangzhou, China (PDF). Transportation Research Record. J. Transport. Res. Board (2011)
33. Vance, S.: Complete streets guidelines codify principles of multi-modal streets. Chicago Streets Blog. 12 April. http://chi.streetsblog.org/2013/04/12/complete-streets-guidelines-codify-principles-of-multi-modal-streets/ (2013)
34. New York Department of Transportation (NY DOT): Complete streets. https://www.dot.ny.gov/programs/completestreets (2013)
35. Calthorpe Associates: Urban design standards for low carbon community. China Sustainable Energy Program, Energy Foundation Report (2010)
36. Russo, R., Grant, B.: Taking it to the (multimodal) streets. The Urbanist. Issue 526. SPUR, San Francisco (2013)
37. Gordon-Koven, L.: Complete streets pay off. 5 November. http://www.smartgrowthamerica.org/2012/11/05/complete-streets-pay-off/ (2012)
38. UK Department of Transport (UK DOT): Low carbon transport: a greener future. http://www.official-documents.gov.uk/document/cm76/7682/7682.pdf (2009)
39. Kennedy, C., Pincetl, S., Bunje, P.: The study of urban metabolism and its applications to urban planning and design. Environ. Pollut. **159**(8–9), 1965–1973 (2011)
40. Chatman, D.G., Noland, R.B.: Transit service, physical agglomeration and productivity in US metropolitan areas. Urban Stud. **51**, 917–937 (2014)
41. Wismans, J., Grahn, M., Denbratt, I.: Low-carbon transport—health and climate benefits. United Nations Centre for Regional Development. Background paper for the Intergovernmental Ninth Regional Environmentally Sustainable Transport (EST) Forum in Asia, November 2015 (2016)

Stephanie Ohshita is Associate Professor of Environmental Science, Studies, and Management at the University of San Francisco, and Visiting Faculty in the China Energy Group, Lawrence Berkeley National Laboratory. Dr. Ohshita's work centers on energy-based strategies for multiple environmental problems—from local air pollution to global climate change—utilizing a mix of engineering, political economy, and institutional analysis. Her research spans China, Japan, the United States, California, and international networks of cities, examining low-carbon resilient cities; energy and environmental policy; emission inventories and climate action plans; cooperation; and equity.

Nan Zhou is Scientist and Group Leader of the China Energy Group of Lawrence Berkeley National Laboratory. Since 2010 Dr. Zhou has worked with the US-China Clean Energy Center-Building Energy Efficiency (CERC-BEE), and in October 2012 she assumed the Director position for CERC-BEE. Dr. Zhou's research has focused on modeling and evaluating China's low-carbon development strategies, assessing building energy efficiency policies and technologies, and development and evaluation of China's appliance standards and labeling program.

Low-Carbon Waste Management

Eugene Mohareb and Daniel Hoornweg

Abstract Waste management is a significant source of urban GHG emissions, with inventories suggesting that it contributes 5 % of the total, on average. Landfills are the dominant source of urban waste GHG emissions, due to their production of methane from the degradation of organic waste. A number of management strategies (e.g., LFG collection, oxidizing covering materials) can be implemented to reduce emissions from landfill operations. Organic waste can also be diverted to other treatment options (composting, anaerobic digesters) that reduce both direct process emissions and indirect emissions through the use of coproducts such as energy and soil amendments. Thermal management practices provide co-benefits such as improved material recovery and energy services, while studies of health implications have generally been inconclusive or have demonstrated no convincing evidence to directly link these treatment approaches with health outcomes. "Three R" approaches to waste management have additional benefits outside the recovery of valuable materials (e.g., aluminum, steel), in that they also can provide a significant indirect emission savings. Further to waste management infrastructure, systemic approaches, such as extended producer responsibility and product service systems, should be employed to shift waste mitigation incentives from cities to manufacturers towards higher diversion rates.

Keywords 3R (reduce-reuse-recycle) • Integrated waste management • Cities

Key Terms

Direct Emissions: GHG emissions from waste management related to the treatment process for a given waste stream. For example, this can refer to methane emissions associated with landfill operations, carbon dioxide related to the combustion of fossil carbon, and fugitive nitrous oxide or methane from composting operations

E. Mohareb (✉)
School of the Built Environment, University of Reading, Reading, Berkshire, UK
e-mail: e.mohareb@reading.ac.uk

D. Hoornweg
University of Ontario, Oshawa, ON, Canada

S. Dhakal, M. Ruth (eds.), *Creating Low Carbon Cities*,
DOI 10.1007/978-3-319-49730-3_11

and bioreactors. Emissions associated with transportation to waste treatment facilities may also be considered in this classification but are typically quantified within transportation-related emissions.

Indirect Emissions: Indirect or life-cycle GHG emissions include embodied GHGs related to the extraction, processing, or transportation of materials within the municipal waste stream. These become relevant in the conversation around GHGs from waste in the instance where economically recoverable and recyclable/reusable materials are not diverted from landfill or incineration, resulting in additional demand for virgin resources and perpetuating linear systems of consumption.

Biogenic Carbon Emissions: GHG emissions associated with nonfossil carbon stored though photosynthesis. When released through processes such as aerobic degradation (composting, open dumping) or combustion (incineration, open burning), biogenic carbon is suggested to be carbon neutral, since the net change in global warming potential (GWP) is zero (i.e., carbon captured through photosynthesis as carbon dioxide is then being rereleased as carbon dioxide). However, under anaerobic conditions (such as in a sanitary landfill or an anaerobic digester), biogenic carbon emissions include methane, which has a GWP of 34 over a 100-year time frame, when accounting for climate feedbacks (IPCC 2013).

Fossil Carbon Emissions: GHG emissions associated with the release of carbon derived from fossil energy sources, generally through incineration. Since the rate of biodegradation of fossil carbon is relatively slow in landfills, they represent a negligible source of fossil carbon emissions.

Coproducts: A number of coproducts of waste management options have the potential to reduce GHG emissions from other sources. The most common examples include electricity or space heating generated from waste treatment operations that produce energy (incinerators, anaerobic digesters), soil amendments (produced from composting or anaerobic digesters), or supplementary cementitious materials (bottom ash from incinerators).

1 Introduction

Waste management is a relatively small yet ubiquitous source of greenhouse gas (GHG) emissions from cities. These emissions are generally associated with methane released due to the landfilling of biogenic carbon. A number of alternatives to landfill disposal have been increasing in prominence in recent decades, including waste-to-energy approaches, anaerobic bioreactors, composting, and recycling. All waste treatment options can provide useful coproducts in addition to their main function of waste treatment, such as energy generation or demand reduction of virgin materials and inorganic fertilizers, all of which have indirect GHG emission implications that must be considered.

2 Waste and Greenhouse Gas Emissions

Waste management is an important component of urban GHG emissions over which local governments tend to have substantial influence. The most significant direct sources of GHGs from waste management activities include anaerobic decomposition of biogenic carbon in sanitary landfills, combustion of fossil carbon in incinerators, and controlled composting/digestion of organic waste streams. In countries with poor waste collection services, wastes are often discharged to local water courses, which causes significant methane releases (due to anaerobic digestion in anoxic waters), as well as increased local flooding (as storm water drain capacities are blocked by waste) and increased incidences of disease (as vectors like rats and mosquitoes increase, e.g., the Plague in Surat, India [1]). The focus of waste management systems has expanded from mainly controlling local pollution and disease vectors to include the goal of maximizing the potential value of components of the waste stream and the reduction of indirect (upstream and downstream) environmental impacts. Through integrated solid waste management, opportunities to reduce both the direct and indirect emissions associated with waste disposal have been increasingly explored by reframing waste as a resource.

The IPCC states that waste directly contributed to 3.0 % of global GHG emissions in 2010 [1, p. 45]. To gain an appreciation of the scale of indirect emissions, the United States Environmental Protection Agency (USEPA) estimates that 42 % of US GHG emissions in 2006 were related to the production, processing, transportation, and disposal of food and materials [2] (see Fig. 1). Examining urban

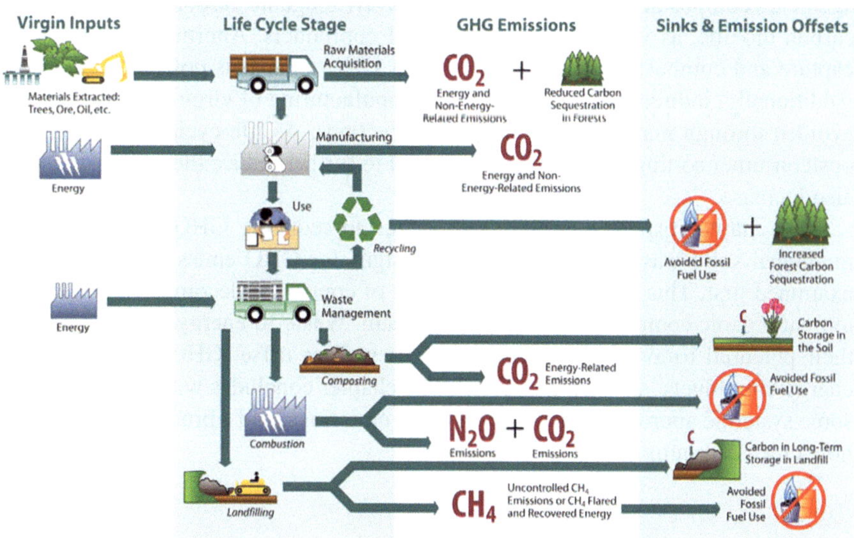

Fig. 1 GHG sources and sinks associated with material life cycles. From [6]

Table 1 Waste fraction of total urban GHG emissions ($n = 44$)

	Waste as a share of urban GHGs (%)
Developing nations	14.7
Developed nations	2.6
Global min	0.4
Global max	41.0
Global average	5.3

Data modified from [36]

emissions from waste management, GHG inventories of over 40 global cities found that between 0.4 and 41 % of urban emissions quantified were attributable to waste, with an average of 5.3 % amongst all cities that were studied (see Table 1 [36]). Landfill sites are the dominant source of GHGs from waste, with Bogner et al. [3] suggesting that they are responsible for half of global waste sector emissions. Methane emissions from landfills, the dominant GHG from landfill operations, are associated with the degradation of biogenic carbon (food scraps, paper and plant trimmings) under anaerobic conditions. Even cities that have attempted to mitigate these landfill gas (LFG) emissions through various waste diversion programs and LFG capture systems have observed waste sector emissions that are dominated by landfill emissions [4].

A multitude of approaches are available to reduce both direct and indirect GHG emissions from waste management. Diverting biogenic materials from landfills provides one means by which direct GHG emissions from waste can be reduced. Methane generation can be avoided through alternative approaches to waste management (composting, incineration) that convert biogenic carbon directly back to carbon dioxide, as well as producing useful coproducts. Alternatively, improved capture and combustion of methane reduces GHG emissions post-decomposition. Additionally, indirect emissions from the manufacturing of virgin materials can be avoided through material recovery (i.e., recycling). As life-cycle accounting and postconsumer sorting improves, the potential to further reduce these emissions will also increase.

This chapter explores current approaches to reducing GHG emissions from municipal solid waste (MSW). Landfill designs for GHG emission reduction are examined first. This is followed by a review of controlled decomposition methods to treat organic components of the waste stream. Waste-to-energy technologies and their potential to avoid direct emissions, as well as offset GHG emissions from energy coproducts, are then examined. The chapter concludes with a discussion of some systemic approaches to reducing GHG emissions, and a broader discussion of how cities can mitigate GHG emissions.

3 Low-Carbon Landfill Design

Landfills receive the greatest share of MSW globally, with Hoornweg and Bhada-Tata estimating that this treatment option handles nearly 350 Mt. annually [7]. The share of waste sent to landfill varies by region; for example, 42 % of waste from OECD countries are sent to landfills, while over 90 % of African waste is deposited in landfills and open dumps. When properly designed and operated, sanitary landfills can provide a waste treatment option that is relatively safe and environmentally benign when compared with the open dumping practices that they have replaced [8]. The control and treatment of leachate through liners and collection systems can reduce subsurface emissions, avoiding the potential contamination of water resources. This captured leachate can also be recirculated to accelerate LFG production, providing improved control over this process and reducing the landfill's operating life. The minimization of GHG emissions can mainly be achieved through LFG capture systems, as well as engineered covering materials that enable the oxidation of the methane component of LFG.

Landfill gas collection systems are typically comprised of a network of vertical gas extraction wells, which are under negative pressure and interspersed within the covered sections of the fill (see Fig. 2). The LFG is drawn through this network, and either flared or upgraded and utilized for energy conversion (such as reciprocating engines for electricity production, combined heat and power plants, or fed into the natural gas grid). It is suggested that these systems can achieve collection efficiencies of up to 90 %, assuming a relatively low rate of decay, as well as aggressive

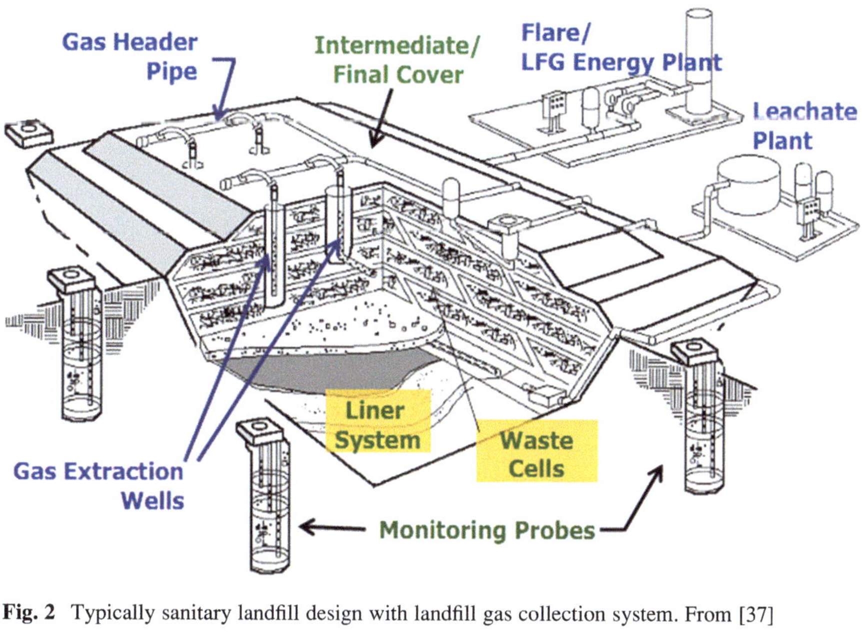

Fig. 2 Typically sanitary landfill design with landfill gas collection system. From [37]

approaches to cover waste and collect LFG [9]. An alternative to LFG capture is to circulate air through the landfill, creating an aerobic environment; biodegradation in an aerobic landfill results in biogenic carbon being transformed back to carbon dioxide, with methane emissions reduced by up to 90 % [5].

Clay landfill-covering materials have previously been used to prevent infiltration of moisture, hence reducing the rate of decomposition of waste. The landfill covers interact with gaseous emissions by hosting methanotrophic communities within the top layer of compost material; these are able to oxidize methane that is not collected through gas capture systems [10]. The oxidation rate of these covering materials ranges from a negligible amount to 100 %, with an average value of 40 % [11]. This rate depends on moisture content, methane concentration, and temperature; decreased moisture and elevated methane concentration limit the oxidation capacity of these microbes, which is greatest at a temperature of 30 °C, with thresholds of roughly −5 °C and 55 °C.

Accelerated decomposition of landfilled organic waste through the provision of nutrients and moisture can be achieved by circulating leachate (bioreactor landfills), which can be complemented with additional water (flushing bioreactor landfills) [8]. While this approach in itself does not reduce GHG emissions, it can accelerate the capture of future emissions, and potentially reduce the time frame of obligations for LFG management. When coupled with a grid-connected energy recovery system, there is also the potential to offset emissions further and assist efforts to reduce the carbon intensity of the electricity grid.

The anaerobic conditions within landfills also provide an opportunity to store carbon; in the absence of oxygen, cellulosic and hemicellulosic components do not completely decompose due to the presence of lignin [12]. As a result, carbon that would have otherwise been released to the atmosphere as carbon dioxide under aerobic conditions remains stored within the landfill. The US EPA [38] estimates that 13 Mt. of equivalent carbon dioxide was stored in US landfills in 2011 from the disposal of yard trimmings and food scraps.

4 Organic Waste Diversion

Composting and anaerobic bioreactors have been used to divert waste from landfills and prevent biogenic methane releases. Composters, in their various incarnations, provide aerobic environments for decomposition in an effort to avoid methane generation. Conversely, anaerobic digesters generate methane in a controlled environment, so that the resultant biogas can be utilized as an energy source.

Composting of MSW can be implemented at varying scales, from centralized industrial-scale facilities (through the collection of source-separated organics, SSO) to home-scale systems. All scales of composting function to reduce the volume of waste and provide a valuable soil amendment coproduct. Some studies have suggested that substantial methane and nitrous oxide emissions (GWP of 34 and 298, respectively [39]) occur regardless of scale of treatment scales; however, avoided emissions from coproducts tend to exceed these quantities [13]. Estimates of net GHG emissions from open composting in windrows, closed composting, and home-scale composting are −37, −32, and −58 kg CO_2e/t of wet waste, respectively [13].

Anaerobic digestion has been used as a method for the stabilization of agricultural and wastewater biosolids. Recently, this approach has been promoted as a means to process organic waste streams of MSW. This generally involves the development of large-scale digestion facilities where SSOs are mixed with other waste streams in order to improve methane yields. Biogas (45–70 % methane) generation can range from 60 to 200 m^3/tonne of wet waste, depending on waste composition [14, 15]. This biogas can be used directly in onsite or offsite combined heat and power applications or further upgraded to be used in other applications (e.g., vehicle fuels, injection into the natural gas grid) [16]. Fugitive emissions from these operations are a concern, with the IPCC 2006 inventorying methodology suggesting a default loss of 5 %, but also suggesting that these are negligible in facilities with systems in place to flare these unintended emissions [40]. Emissions avoided from coproducts (energy and soil amendments) reduce net GHG emissions; in the case of the Greater Toronto Area, a net carbon sink was suggested through consideration of the displacement of electricity grid emissions alone [5]. Alternative energy-generating treatment options to biogas include biohydrogen and bioethanol production, with a number of studies having demonstrated the potential to convert organic wastes to these alternative forms of energy [17, 18].

5 Thermal Treatment of Waste

Thermal treatment is another common approach to deal with waste, from open burning to more modern waste-to-energy (WTE) iterations. Benefits can include lower GHG emissions, smaller land requirements (mass and volume reductions of up to 75 % and 90 %, respectively [19]), generation of electricity (a fraction of this being renewable when considering biogenic carbon), and improved recycling rates [20]. In the case of direct WTE facilities, two primary options provide a means to reduce GHGs from the broader MSW stream. Incineration has historically been a prominent option for waste disposal, but has faced considerable opposition in recent decades due to health concerns (such as those related to the release of dioxins and furans). More recently, gasification has provided an alternative WTE option that addresses many of the shortcomings of incineration (less efficient energy recovery, health concerns associated with dioxin and furan emissions; discussed below) [21]. A key deterrent to WTE and especially gasification is the high costs of operation *vis-à-vis* landfilling. Costs are typically three to four times higher for WTE (including sale of potential energy) and some ten times higher for gasification: and these higher costs generally provide no commensurate environmental benefits [7].

Incineration has the benefit of converting biogenic carbon to carbon dioxide directly, avoiding the methane emissions associated with landfilling, as well as alleviating potential long-term environmental concerns associated with landfill sites. Additionally, useful coproducts including energy, recovered metals, and other components of bottom ash provide additional life-cycle GHG mitigation. A life-cycle analysis of GHG emissions from a typical European incineration operation, which considered upstream emissions from inputs and downstream emissions offset by energy products, suggested the following ranges of emissions [22]:

- Upstream emissions of 59 to 158 or 7 to 62 kg CO_2e per tonne of wet waste for high- or low-grid carbon intensity, respectively
- Direct emissions of 347 to 371 kg CO_2e per tonne of wet waste, dominated by fossil carbon components of the waste stream and fossil fuels used in the combustion process
- Downstream emissions of −811 to −1373 or −480 to −712 kg CO_2e per tonne of wet waste for high- or low-grid carbon intensity, respectively

To clarify, these suggest that in a region with an electricity grid that has a high carbon intensity per kWh generated, there is a greater net climate benefit from incinerating waste; however, given the current global trend towards decarbonizing grids, it follows that incineration will become a less attractive option from a GHG perspective. It is relevant to highlight that recycling rates observed in US jurisdictions with WTE programs are comparable to the national average, suggesting complementary reductions in indirect emissions [20].

Health concerns have created public resistance to incineration, with many studies finding increased risk of non-Hodgkin's lymphoma and sarcomas associated with incineration emissions [23]; however, Giusti emphasizes that many of these studies examine older incineration facilities and that food consumption tends to be the intake pathway, not inhalation. Studies from France found a weak link between non-Hodgkin's lymphoma and exposure to dioxins, with authors speculating causation from other pollutants, as well as outdated incineration technology. As well, Giusti reports that studies in the USA and the UK did not find conclusive evidence of causation between proximity to incineration and the health impacts they examined [23].

Similarly, with gasification of waste, the provision of heat and electricity (from the utilization of the secondary energy product converted from solid waste) has the potential to offset other, more carbon-intensive energy demand [21]. In the case of gasification, the synthesis gas generated from the process can be stored and/or used offsite in any suitable (and potentially more economically and energetically efficient) application that is desired. Additionally, it is suggested that gasification provides a number of benefits beyond those from conventional WTE systems, including potentially lower emission control costs, lower emissions of dioxins and furans, and ability to recover metals in a non-oxidized from.

In low-income countries, low-temperature, open burning of waste is common. This contributes significant local particulate air pollution, with appreciable negative health impacts, and black carbon, an important short-term climate forcer. Efforts are currently under way to curtail short-lived climate pollutants from open burning through United Nation's Climate and Clean Air Coalition [24].

6 Source Reduction, Reuse, Recycling, and Recovery

The scale of raw material extraction, production of goods, and distribution of final products, coupled with intermediate transportation, suggests that emissions associated with activities that are upstream of consumption are likely to be more resource/carbon intensive than those downstream through waste management activities [25]. This is indeed suggested by the US estimate that all emissions associated with the provision of materials and food amounted to 42 % of the total in 2006 when compared in scale to approximately 2 % of total emissions that was contributed by waste in the same year [3, 26]. Therefore, it is prudent to focus waste management efforts on reuse, source reduction, and, perhaps to a lesser extent, recycling in order to achieve deeper life-cycle GHG mitigation.

Waste management approaches to source reduction, reuse of waste, and recycling avoid the extensive upstream emissions associated with many of the essential materials consumed in urban economies. Source reduction prevents the development of resources and production of consumer materials altogether, while reuse efforts extend their service life with minimal maintenance. Energy inputs required to recycle materials diminish the net energy benefits; however, this treatment option still generally proves to be worthwhile from a climate perspective (is presented in Table 2).

Recycling can follow either an open-loop or closed-loop path; closed-loop recycling suggests that the material is used in the same process continuously, whereas open-loop recycling involves the material being used in a product that is distinct from its original purpose [41]. The life-cycle energy and GHG emission

Table 2 GHG emission estimates in kilograms of equivalent carbon dioxide (kg CO_2e) associated with the recycling of various materials

Waste material recycled	Jurisdiction	GHG emissions (negative value implies avoided emissions)	Source
Plastic	Europe	−1500 ↔ −700 kg CO_2e/t wet weight (virgin plastic substitution) −1200 ↔50 kg CO_2e/t wet weight (incineration-energy substitution)	[27]
Aluminum	Europe	−19,300 ↔ −5000 kg CO_2e/t wet weight (primary aluminum substitution)	[28]
Aluminum	Various Europe, USA	−13,500 ↔ −9200 kg CO_2e/t wet weight (primary aluminum substitution)	[28]
Steel	Europe	−2400 ↔ −600 kg CO_2e/t wet weight (primary steel substitution)	[28]
Steel	Various Europe, USA	−1800 ↔ −700 kg CO_2e/t wet weight (primary steel substitution)	[28]
Paper	Europe	−4400 ↔ 1500 kg CO_2e/t wet weight (substitution of virgin forestry products)	[29]
Paper	Europe, USA	−3900 ↔ 200 kg CO_2e/t wet weight (substitution of virgin forestry products)	[29]

impacts related to the finer details of either path must be carefully considered in any analysis of the benefits of a recycling system, as suggested by the broad ranges presented in Table 2. It is also worth considering that losses do occur in recycling processes, due to contamination and inefficiencies within the recovery system; for example, loss estimates for steel, aluminum, plastic, and paper are suggested to be up to 1–5 %, 2–10 %, 3–10 %, and 2–18 % from the point of separation from the waste stream until it is prepared for reuse [27–29]. For a more complete understanding of the GHG implications of urban waste management options, the USEPA Waste Reduction Model (WARM) enables municipalities to quantify life-cycle emissions associated with recycling and source reduction, providing insight into the potential emission reductions achieved through these approaches relative to other waste management options [41].

Though recycling from diverted waste is the most common pathway to recover valuable materials from the waste stream in high-income countries [7], landfill mining (also referred to as landfill reclamation) is an emerging approach. While the focus of many landfill mining operations has been to reclaim space in order to extend the operating life of the landfill, the value of reclaimed materials and co-benefits associated with these has the potential to strengthen the business case for these types of projects, especially when relevant commodity prices are high [30]. Comparing landfills to thermal alternatives in this context emphasizes another advantage over the latter; materials are preserved in their postconsumer state and leaving the potential for future recovery, if reclamation technologies and commodity prices further improve the financial incentive to do so.

7 Systemic Changes Towards GHG Mitigation from Waste

The GHG mitigation approaches described above all assume a linear throughput of resources, where the consumer/municipality makes the ultimate decision with respect to waste disposition. However, systemic changes have been proposed that incentivize either a low-waste or no-waste approach, placing a greater onus of waste management on material use upstream of the consumer. For example, several product-service system models allow the consumer to purchase a service, with ownership of the service delivery mechanism (i.e., the material good itself) remaining with the manufacturer or other service provider [31]. This can eliminate a split incentive that exists in the conventional goods purchasing model, as manufacturers would be able to reap financial gains for providing services that have low maintenance and low operating costs, e.g., disincentivizing "planned obsolescence."

Another approach to promote a lower waste system is through legislating extended producer responsibility (EPR). EPR systems retain the conventional product purchasing model, but extend the responsibility for (and cost of) end-of-life management to the producer, rather than the consumer (or, more directly, the municipality). This also has the effect of incentivizing manufacturers to design for recyclability, as has been conceived in e-waste regulations [32]. EPR legislation have existed in Europe since the 1990s, and are currently being implemented globally [33]. Methods such as these can contribute to broader dematerialization of the global economy.

8 Mitigation at the Urban Scale

The GHG mitigation options discussed above require varying degrees of end-user behavior change, as well as investments in new infrastructure, to ensure their success. Recycling, composting, and anaerobic digestion require both separation (or better stated, "segregation") by the consumer (with recycling occasionally divided into multiple streams, though this is becoming less common) and investment in new infrastructure, such as composting or material recovery facilities (not to mention markets for which to sell recovered materials). Conversely, WTE technologies can reduce the need for source separation by the consumer (useful with residents where source separation is challenging, such as in ill-equipped multiunit residential buildings). However, the impact on the value of coproducts associated with the transformation inherent in WTE processes relative to other diversion alternatives is unclear, especially in the context of improving recycling and separation technologies. Alternative systemic changes such as product-service models or EPR can preempt the need for behavioral and infrastructural changes through upstream initiatives. As discussed above, each of these options provides direct and/or indirect energy benefits, which also need to be considered.

Municipal recycling and resource recovery programs are susceptible to the vagaries of commodities markets. For example, revenues from the sale of recycled materials can fluctuate widely—occasionally reaching negative values, requiring municipalities to pay to dispose of recycled materials. Ultimately, city-specific considerations including, but not limited to, site availability, public acceptance, political will, and access to markets for diverted materials (compost, recyclables, digestate) will all factor into the selection of low-carbon waste treatment alternatives.

To reiterate, waste represents a considerable share of municipal GHG emissions and decisions related to emission mitigation are often directly within the purview of local government. This can simplify approaches to mitigation when compared to the buildings or transportation sectors, where decisions rest with a large number of discrete, decentralized actors. Waste managers have achieved some success to this point in reducing these emissions through the diversion of waste from landfill and by installing LFG collection in existing sites. For example, the City of Toronto, Canada, reduced its GHG emissions from waste by nearly 60 % (2.1 Mt) between 1990 and 2013, through LFG capture and increasing the share of waste treated through anaerobic digestion, composting, and recycling [34]. Diversion improvements, which also have direct and lifecycle GHG implications, have been achieved; between the years 1995 and 2009, the EU-27 nations reduced annual per capita landfill waste by 35 %, while increasing per capita waste incinerated, recycled, and

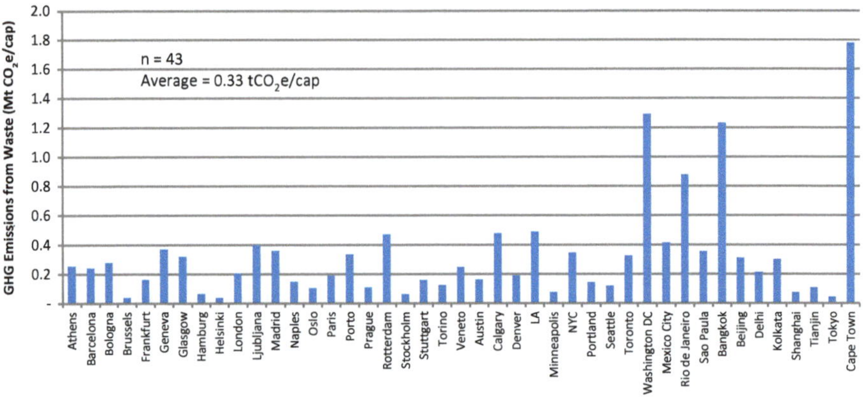

Fig. 3 Per capita GHG emissions from a sample of global urban inventories. Data from [36], no permission needed

composted by 56, 159, 366 and 224 % [42]. Additionally, one-third of US landfills have operational LFG collection systems, with the potential to increase this to nearly 60 % if all candidate sites identified are pursued [43, 44].

Greenhouse gas emissions associated with MSW reported by Kennedy and others [36] ranged from 0.04 to 1.78 t CO_2e per capita, with an average of 0.33 t CO_2e per capita (Fig. 3). Generally speaking, deep reductions are achievable by diverting office paper and food/garden waste from landfill, considering the higher methane yield from these waste components [41].

It is important to highlight that some waste emission inventories do not include waste from the institutional, commercial, and industrial sectors, whose waste production often occurs within urban boundaries and can be substantial (as can the life-cycle GHG emissions related to this waste). For example, 2010 EU-27 waste from construction, services, and manufacturing was nearly six times that from the residential sector (dominated by construction sector waste) [42]. Similarly, nearly two-thirds of waste produced in Canada in 2010 is attributable to the non-residential waste [45]. It is essential that cities make efforts to include these emissions in their inventories, as suggested by the Global Protocol for Cities [35].

References

1. Wilson, D.C.: Development drivers for waste management. Waste Manag. Res. [Internet]. **25**(3), 198–207 (2007). http://wmr.sagepub.com/cgi/doi/10.1177/0734242X07079149
2. U.S. Environmental Protection Agency (EPA): Opportunities to Reduce Greenhouse Gas Emissions Through Materials and Land Management Practices. Washington, DC
3. Bogner, J., Hashimoto, S., Diaz, C., et al.: Mitigation of global greenhouse gas emissions from waste: conclusions and strategies from the intergovernmental panel on climate change (IPCC)

fourth assessment report. working group III (mitigation). Waste Manag. Res. [Internet]. **26**(1), 11–32 (2008). http://dx.doi.org/10.1177/0734242X07088433
4. Mohareb, E.A., MacLean, H.L., Kennedy, C.A.: Greenhouse gas emissions from waste management—assessment of quantification methods. J. Air Waste Manage. Assoc. [Internet]. **61**(5), 480–493 (2011). http://dx.doi.org/10.3155/1047-3289.61.5.480
5. US Environmental Protection Agency: Solid waste management and greenhouse gases—a life cycle assessment of emissions and sinks.
6. Hoornweg, D., Bhada-Tata, P.: What a waste: a global review of solid waste management [Internet]. Washington, DC. http://siteresources.worldbank.org/INTURBANDEVELOPMENT/Resources/336387-1334852610766/What_a_Waste2012_Final.pdf
7. Manfredi, S., Christensen, T.H.: Environmental assessment of solid waste landfilling technologies by means of LCA-modeling. Waste Manag. [Internet]. **29**(1), 32–43 (2009). http://dx.doi.org/10.1016/j.wasman.2008.02.021
8. Barlaz, M., Chanton, J., Green, R.: Controls on landfill gas collection efficiency: instantaneous and lifetime performance. J. Air Waste Manag. Assoc. [Internet]. **59**, 1399–1404 (2009). http://www.tandfonline.com/doi/abs/10.3155/1047-3289.59.12.1399
9. Lou, X.F., Nair, J.: The impact of landfilling and composting on greenhouse gas emissions—a review. Bioresour. Technol. **100**(16), 3792–3798 (2009)
10. Spokas, K.A., Bogner, J.E.: Limits and dynamics of methane oxidation in landfill cover soils. Waste Manag. [Internet]. **31**(5), 823–832 (2011). http://dx.doi.org/10.1016/j.wasman.2009.12.018
11. Barlaz, M.A.: Carbon storage during biodegradation of municipal solid waste components in laboratory-scale landfills. Global Biogeochem. Cycles. **12**(2), 373–380 (1998)
12. Boldrin, A., Andersen, J.K., Møller, J., Christensen, T.H., Favoino, E.: Composting and compost utilization: accounting of greenhouse gases and global warming contributions. Waste Manag. Res. [Internet]. **27**(8), 800–812 (2009). http://wmr.sagepub.com/content/27/8/800.abstract
13. Goldstein, N.: Source-separated organics as feedstock for digesters—the Toronto experience. Biocycle [Internet]. **46**(8), 42–46 (2005). http://www.biocycle.net/2005/08/19/source-separated-organics-as-feedstock-for-digesters/
14. Li, Y., Park, S.Y., Zhu, J.: Solid-state anaerobic digestion for methane production from organic waste. Renew. Sustain. Energy Rev. [Internet]. **15**(1), 821–826 (2011). http://www.sciencedirect.com/science/article/pii/S1364032110002224
15. Holm-Nielsen, J.B., Al Seadi, T., Oleskowicz-Popiel, P.: The future of anaerobic digestion and biogas utilization. Bioresour. Technol. [Internet]. **100**(22), 5478–5484 (2009). http://www.ncbi.nlm.nih.gov/pubmed/19217772
16. Yasin, N.H.M., Mumtaz, T., Hassan, M.A., Abd Rahman, N.: Food waste and food processing waste for biohydrogen production: a review. J. Environ. Manage. [Internet]. **130**, 375–85 (2013). http://www.sciencedirect.com/science/article/pii/S0301479713006002
17. Li, A., Antizar-Ladislao, B., Khraisheh, M.: Bioconversion of municipal solid waste to glucose for bio-ethanol production. Bioprocess Biosyst. Eng. [Internet]. **30**(3), 189–96 (2007). http://www.ncbi.nlm.nih.gov/pubmed/17458580
18. Cheng, H., Hu, Y.: Municipal solid waste (MSW) as a renewable source of energy: current and future practices in China. Bioresour. Technol. [Internet]. **101**(11), 3816–3824 (2010). http://www.ncbi.nlm.nih.gov/pubmed/20137912
19. Psomopoulos, C.S., Bourka, A., Themelis, N.J.: Waste-to-energy: a review of the status and benefits in USA. Waste Manag. [Internet]. **29**(5), 1718–1724 (2009). http://dx.doi.org/10.1016/j.wasman.2008.11.020
20. Arena, U.: Process and technological aspects of municipal solid waste gasification: a review. Waste Manag. [Internet]. **32**(4), 625–639 (2012). http://www.sciencedirect.com/science/article/pii/S0956053X11004314
21. Astrup, T., Møller, J., Fruergaard, T.: Incineration and co-combustion of waste: accounting of greenhouse gases and global warming contributions. Waste Manag. Res. [Internet]. **27**(8), 789–799 (2009). http://wmr.sagepub.com/content/27/8/789.short

22. Giusti, L.: A review of waste management practices and their impact on human health. Waste Manag. [Internet]. **29**(8), 2227–2239 (2009). http://www.sciencedirect.com/science/article/pii/S0956053X09001275
23. UNEP.: UN Climate Summit commitments to reduce short-lived climate pollutants, and their impacts [Internet]. (2014). http://www.unep.org/ccac/Events/UNClimateSummit2014/ActionStatementSupport/tabid/794296/Default.aspx
24. European Environment Agency: Waste opportunities—past and future climate benefits from better municipal waste management in Europe. Copenhagen
25. US Environmental Protection Agency: Inventory of US GHG Emissions and Sinks (1990–2008)
26. Astrup, T., Fruergaard, T., Christensen, T.H.: Recycling of plastic: accounting of greenhouse gases and global warming contributions. Waste Manag. Res. [Internet]. **27**(8), 763–772 (2009). http://www.ncbi.nlm.nih.gov/pubmed/19748943
27. Damgaard, A., Larsen, A.W., Christensen, T.H.: Recycling of metals: accounting of greenhouse gases and global warming contributions. Waste Manag. Res. [Internet]. **27**(8), 773–780 (2009). http://wmr.sagepub.com/content/27/8/773.short
28. Merrild, H., Damgaard, A., Christensen, T.H.: Recycling of paper: accounting of greenhouse gases and global warming contributions. Waste Manag. Res. [Internet]. **27**(8): 746–753 (2009). http://wmr.sagepub.com/content/27/8/746.short
29. Frändegård, P., Krook, J., Svensson, N., Eklund, M.: A novel approach for environmental evaluation of landfill mining. J. Clean. Prod. [Internet]. **55**, 24–34 (2013). http://www.sciencedirect.com/science/article/pii/S0959652612002788
30. Tukker, A.: Eight types of product-service system: eight ways to sustainability? Experiences from suspronet. Bus. Strateg. Environ. **13**(4), 246–260 (2004)
31. Huisman, J.: Too big to fail, too academic to function. J. Ind. Ecol. [Internet]. **17**(2), 172–174 (2013). http://doi.wiley.com/10.1111/jiec.12012
32. Mayers, C.K.: Strategic, financial, and design implications of extended producer. J. Ind. Ecol. **11**(3), 113–131 (2007)
33. City of Toronto. Toronto's 2013 Greenhouse Gas Inventory [Internet]. Toronto. http://www1.toronto.ca/wps/portal/contentonly?vgnextoid=44a0e211597a1410VgnVCM10000071d60f89RCRD
34. Fong, W.K., Sotos, M., Doust, M., Schultz, S., Marques, A., Deng-Beck, C.: Global Protocol for Gas Emission Inventories—An Accounting and Reporting Standard for Cities. World Resources Institute, Washington, DC

Internet References

35. Kennedy, C.A., Ramaswami, A., Carney, S., Dhakal, S.: Greenhouse Gas Emissions Baselines for Global Cities and Metropolitan Regions. http://siteresources.worldbank.org/INTURBANDEVELOPMENT/Resources/336387-1256566800920/6505269-1268260567624/KennedyComm.pdf (2009). Accessed 23 Nov 2013
36. USEPA: An overview of landfill gas energy in the United States. http://www.epa.gov/lmop/documents/pdfs/overview.pdf (2013). Accessed 8 Apr 2014
37. USEPA: Inventory of US Greenhouse Gas Emissions and Sinks: 1990–2011. http://www.epa.gov/climatechange/ghgemissions/usinventoryreport.html (2013). Accessed 17 Nov 2013
38. Intergovernmental Panel on Climate Change: Working Group I Contribution to the Fifth Assessment Report: Climate Change 2013—The Physical Science Basis, Final Draft Underlying Scientific-Technical Assessment. http://www.climatechange2013.org/images/uploads/WGIAR5_WGI-12Doc2b_FinalDraft_All.pdf (2013)
39. Intergovernmental Panel on Climate Change: Guidelines for National Greenhouse Gas Inventories. http://www.ipcc-nggip.iges.or.jp/public/2006gl/ (2006). Accessed 17 Nov 2013

40. USEPA: Solid Waste Management and Greenhouse Gases. http://epa.gov/epawaste/conserve/tools/warm/SWMGHGreport.html (2013). Accessed 25 Nov 2013
41. Eurostat: Environment Tables. http://epp.eurostat.ec.europa.eu/portal/page/portal/environment/data/main_tables (2011). Accessed 9 Apr 2014
42. US EPA: Energy Projects and Candidate Landfill. http://www.epa.gov/lmop/projects-candidates/index.html (2014). Accessed 9 Apr 2014
43. US EPA: Municipal Solid Waste Generation, Recycling, and Disposal in the United States—Tables and Figures for 2012. http://www.epa.gov/waste/nonhaz/municipal/pubs/2012_msw_dat_tbls.pdf (2014). Accessed 9 Apr 2014
44. City of Toronto: Summary of Toronto's 2011 GHG and Air Quality Pollutant Emissions Inventory. http://www.toronto.ca/legdocs/mmis/2013/pe/bgrd/backgroundfile-57187.pdf (2013). Accessed 9 Apr 2014
45. Statistics Canada: Waste Management Industry Survey—Business and Government Sectors, 2010. http://www.statcan.gc.ca/pub/16f0023x/16f0023x2013001-eng.pdf (2013). Accessed 9 Apr 2014

Eugene Mohareb is a Lecturer in Sustainable Urban Systems in the School of the Built Environment at the University of Reading. His research focuses on urban greenhouse gas (GHG) source and sink quantification, implications of technology change scenarios on future GHG emissions, as well as urban approaches to climate change mitigation.

Daniel Hoornweg is the Richard Marceau Chair at the University of Ontario Institute of Technology. He researches sustainable cities—public policy and efficiencies of urban metabolism. Dan has some 25 years' solid waste experience working in cities as waste management coordinator and with cities as urban advisor with the World Bank.

Managing Greenhouse Gas Emissions in Cities: The Role of Inventories and Mitigation Action Planning

Flavia Carloni and Vivien Green

Abstract The three main questions that must be addressed in a way to make cities' GHG emission inventories comparable are how to draw the borders, what to measure, and how to measure. Objectives of monitoring and quantifying the GHG effects of mitigation actions vary across cities. The City of Rio de Janeiro is a good example of a city that is taking action to institutionalize the climate change issue, with the Municipal Policy on Climate Change and the Low Carbon City Development Program. This chapter provides an overview of experiences and best practices to help cities design their own roadmap to a climate mitigation policy. It is important to allow comparability among emissions from different studies from different cities and help promote cooperation among them in mitigation and adaptation to climate change.

Keywords GHG emission inventory • Climate change mitigation • Cities • Rio's Low Carbon City Development Program • Cities' best practice

1 Introduction

The objective of this chapter is to identify opportunities and demonstrate the feasibility of adopting local initiatives in managing and controlling GHG emissions. For this purpose, a review of the state of the art of international experiences in cities worldwide is undertaken. The authors report on the wealth of knowledge, initiatives, and ongoing work showing that although most experiences have been significant, some have been conducted applying different methodologies and/or protocols, implying a lack of standardization and comparability among different

F. Carloni
Center for Integrated Studies on Climate, Federal University of Rio de Janeiro, Rua Roberto Rowley Mendes, 24210-310 Rio de Janeiro, Brazil
e-mail: flaviabbac@gmail.com

V. Green (✉)
Centro Clima Laboratory, Federal University of Rio de Janeiro, Avenida Roberto Silveira, 18/306—Icarai, 24230-165 Rio de Janeiro, Brazil
e-mail: vivien.green@gmail.com

S. Dhakal, M. Ruth (eds.), *Creating Low Carbon Cities*,
DOI 10.1007/978-3-319-49730-3_12

sites. Some recommendations to face these challenges are made, based on experiences from the city of Rio de Janeiro, Brazil.

Cities may be seen as a planning unit for mitigation management purposes [1]. Given their specificities and similarities, individual experiences can be shared to define a common framework to improve mitigation actions.

2 Ancillary Effects of Climate Policies at the Local Level

The greenhouse gas (GHG) emissions of a city, region, or country arise from the burning of fossil fuels (oil, natural gas, and coal), waste treatment, industrial processes, and changes in land use, among others. Virtually all economic sectors of modern society (industry, services, transport, agriculture, and construction) produce carbon dioxide (CO_2), methane (CH_4), and nitrous oxide (N_2O), the main greenhouse gases. GHG emission estimates have an inherent uncertainty due to the difficulty in obtaining data on all of these activities and emission factors. This is even more challenging when dealing with cities where delimiting the boundaries of activities is more complex [1].

However, some cities' governments view efforts to reduce GHG emissions as jeopardizing their economic and social agendas [2], once the economic development and population growth induce an increase in GHG emission. Yet it is possible to combine those goals, once there is paths to promote development and climate agendas in ways that result in positive outcomes for both.

For example, the same combustion process that causes GHG emissions also generates conventional pollutants with adverse effects on human health, ecosystems, agricultural productivity, and materials [2]. Therefore, a greenhouse gas mitigation strategy creates positive effects for public finances due to the prevention of damages related to local pollution. The literature provides many examples of similar such co-benefit opportunities[1] [3–6]. Moreover, due to the carbon market, financial resources from GHG mitigation can ultimately benefit cities.

Accounting and Mitigating GHG Emissions

As pointed out by Gurney et al. [7], while different methods to account for community-scale emissions have been designed by various organizations—such as the World Business Council for Sustainable Development and the World Resources Institute—it observes that most cities around the world didn't establish an independent, comprehensive, and comparable sources of data. As described by the authors, these can refer not only with the cost to produce an emission report but also with the lack of expertise. The authors also describe how the transparency of

[1]Some trade-offs can occur and therefore an integrated assessment could avoid these co-costs.

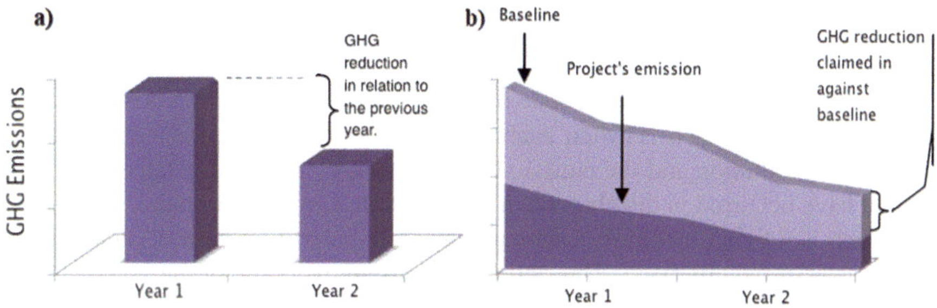

Fig. 1 Quantification of GHG emission reduction based on the analysis of an inventory (**a**) and based on the analyses of a mitigation action (**b**). Translated from [1] with permission

data and methods is crucial to develop the inventory, raise trust, and in the end enable the establishment of verification by third parties.

Furthermore, accounting methodologies differ between those that track the global emissions of a city (the city's GHG inventory) and those that monitor the mitigation of particular actions. The main methodological difference between GHG inventories and accounting for the emission reductions of a particular action is that the first compares the absolute emissions in 1 year with respect to another year, and the second compares the emissions' decrease relative to a baseline scenario that considers the absence of the action (Fig. 1) [1].

There are two possible monitoring approaches:

1. A comparison of two or more emission inventory data sets (total values or sectorial and sub-sectorial values): In this case, it is possible to observe the historical trend or to set targets, comparing one or more years to a baseline year
2. Scenarios building techniques to assess the mitigation outcomes of specific policies, projects, and measures

Both monitoring systems are useful to assess past and future performances of the city, the former being a snapshot and the latter an analytical tool that allows the understanding of the consequences of specific mitigation measures.

Among the challenges faced to apply these approaches in cities, some are highlighted below:

- Whether and how to account for indirect emissions (leakages[2])
- How to ensure additionality[3] and reduction of GHG emissions by investment in projects or through the purchase of credits generated

---

[2]Carbon leakage is defined as the increase in emissions outside a region as a direct result of the policy to cap emission in this region. Carbon leakage means that the domestic climate mitigation policy is less effective and more costly in containing emission levels, a legitimate concern for policy makers.

[3]Additionality is the requirement that the greenhouse gas emissions after implementation of a CDM project activity are lower than those that would have occurred in the most plausible alternative scenario to the implementation of the CDM project activity.

- Whether and how to demonstrate cause-and-effect relationships between a given action and a given emission reduction [1]

Also, this monitoring can be done by taking a bottom-up or a top-down strategy. A bottom-up approach is when each mitigation intervention is developed as an individual mediation and the emission reductions are calculated relatively to what would have occurred in the absence of intervention. On the other hand, a top-down approach is a GHG inventory calculated and compared to a case designed in the business-as-usual approach. This approach gives an overview of the scope and the target. However, the major drawback is that designing the business-as-usual emissions of an entire city is hard and varies, considerably, among different conditions—such as population growth, economic development, and expansion of the major emitting sectors.

It is a challenge to elaborate an inventory using one of those approaches but a bigger challenge is to combine the top-down and bottom-up methods. But more and more scientists are noticing how important it is to address both procedures and are trying to overcome the challenge to put those together [7].

Also there is a large discussion about GHG inventories and their uncertainties, how they are conducted, methods used, and how they deal with these uncertainties and report a fair estimate of the emissions. Ometto et al. [8] address that accounting emissions involves different uncertainties due a variety of reasons such as the lack of availability of sufficient and appropriate data and the techniques for processing them.

Box 1

Ometto et al. [8] discussed in their work six insights that can be taken to better address and minimize these uncertainties and should require further attention.

1. Verification: reconciling bottom-up and top-down GHG emission analyses
2. Avoiding systemic surprises: distinguishing between subsystems with fundamentally different emission-dynamic and uncertainty characteristics before superimposing them
3. Making uncertainty analysis a key component of national GHG inventory analysis to support the development of informed policy in the framing of international environmental agreements: providing advanced guidance, beyond the methodologies offered by the IPCC, to ensure that uncertainty is dealt with appropriately in an internationally consistent way across countries, subsystems, sources and sinks, GHGs, and sectors
4. Minimizing the impact of uncertainty to support the design of advanced policy agreements: providing approaches that allow subsystems to be treated individually and differently rather than collectively (in terms of CO_2 equivalence) and equally (not distinguishing between emissions and removals)

(continued)

Box 1 (continued)

5. Full GHG accounting: ensuring that any differentiated approach to accounting forms a logical subset of a full GHG accounting approach
6. Compliance versus reporting (bifurcation of agreements) but in a complementary manner: providing options that allow for smarter treatment of subsystems, for example, individually and differently, while at the same time following full GHG accounting

Besides the choice regarding the methods they must follow a step-by-step methodology to ensure both a good analysis during the process and the possibility of future comparison among different inventories—thus a better understanding of the uncertainties.

3 International Practice: GHG Inventories in Cities

Main Methodologies

In 2006, the International Panel for Climate Change (IPCC) published the IPCC Guidelines for National Greenhouse Gas Inventories [9], which is frequently adapted for cities' GHG inventories. Three years later, the International Council for Local Environmental Initiatives (ICLEI) issued the International Local Government GHG Emissions Analysis Protocol (IEAP 2009) [10] as a first attempt to provide a more adjusted outline for local GHG inventories, with the support of a software named HEAT. Subsequently, the World Bank, UNEP, and UN-Habitat developed the International Standard for Determining Greenhouse Gas Emissions for Cities (2010). Recently, these institutions joined efforts and started working on the Global Protocol for Community-scale Greenhouse Gas Emissions (2011) with the objective of providing a common format for GHG emission accounting and reporting within cities.

The United States Environmental Protection Agency [11] brings in its website a very simple and direct guide to conduct a GHG inventory: (1) set the boundaries—being either physical, operational, or governmental; (2) define the scope—considering which emission sources should be included in the report, and also which gases are going to be investigated; (3) choose quantification approach—consider the data availability and the purpose of the inventory to adopt either a top-down, bottom-up, or hybrid approach; (4) set the baseline—determine the baseline year according to data availability and representativeness to be used as a benchmark to monitor progress and allow comparison among different years; (5) engage stakeholders—bring the stakeholders into the process in the very beginning with the intention to collect more data and information and help construct a public acceptance; and

(6) consider certification—a third-party review and certification of the methods and data is highly advised in this way assuring high quality, consistency, and transparency of the report.

Despite the progress made in the last years, and the different methodologies, there are still some questions for addressing GHG accounting approaches at local scales:

How to Draw Boundaries for Analysis?

System boundary definitions present a difficult issue because cities are polar centers that concentrate commerce and services, attracting people from surrounding areas on a regular basis. One way of defining the inventory boundaries is to use the city's territory limits; another is to refer solely to the public sector-related emissions; yet another is to relate to the consumption or production activities of the city's inhabitants. Each of these system boundaries holds complications and needs further decisions over accounting methods. A recurrent problem is the double counting, since a wide set of data is needed and different contributing institutions sometimes account for overlapping activities.

Dubeux [12] based on the first inventories made for the cities of Rio de Janeiro and São Paulo—in the early 2000s—provides a series of examples of boundary delimitation difficulties in cities, focusing those present in metropolitan regions. These have remained unsolved due to the lack of adoption of a unique protocol, until the launch of the Global Protocol for Community-Scale Greenhouse Gas Emission Inventories (GPC), in 2014, that intends to promote a consistent and transparent measurement and reporting of greenhouse gas emissions among cities.

What to Measure?

On the measurement issue, some consensus has been built in recent years: city inventories are generally presented in terms of individual gases as well as in CO_2-equivalent units. Noteworthy is the fact that information about all the gases required by IPCC Guidelines is not easy to obtain and, thus, several municipalities currently only consider a subset of them (e.g., CO_2, CH_4, and N_2O).

A frequent division of scopes considers scope 1, the direct emissions; scope 2, the imported energy-related emissions, as in the cases of electricity imports from the grid; and scope 3, emissions due to the life cycle of goods consumption, e.g., agricultural related emissions. Aside from the evident complexity of calculating life cycle-related GHG emissions, one must also pay attention to emission factors from the grid (which often take into account electricity producers that reside in the city) and similar problems. For example, the city of Rio de Janeiro imports around 85 % of its total electricity consumption. Therefore, in addition to the GHG emissions from the burning of fossil fuels to generate electricity within the city, the emissions from the imported electricity are also included, which are calculated using the emission factor of the National Interconnected System grid, according to UNFCCC ACM 0002 methodology. To calculate the emission factor for the electricity generated inside the city of Rio de Janeiro, data from the existing energy supply mix of the city was used.

The level of sectorial detail in which to report also poses challenges to cities' inventories. The IPCC recommends five sectors [9]: energy; industrial processes; agriculture, forests, and land use; waste; and others.[4] The energy sector is further subdivided because, normally, it encompasses a great variety of activities. However, this division may not suit the interests of some cities that may need more detail in some particular sector to track its progress.

How to Measure?

Most methodologies calculate emissions through activity data and related emission factors.[5] These factors may be local, national, or even international default values. Activity information is, generally, available—although sometimes only for part of the emitting activities in the city. Thus, one may need to estimate these figures through interpolation or other method projections, or even report that some GHG sources are not considered. Emission factors are a more problematic topic, since calculating all local values poses a tough task for inventories. Therefore, many cities turn to national or international references.

There is also the matter of data quality. Carloni provides a review of some best practices in this context [1]. Ranganathan [13] also highlights that the quality management of data needs to be part of a full inventory program, ensuring engagement with the following principles: relevance, completeness, consistency, transparency, and accuracy.

Practical Experiences

Inventory Accounting

Table 1 presents some of the tools used by cities to elaborate GHG inventories.

There are significant differences in boundary setting (Table 2) and reporting sectors (Table 3) since each tool uses diverse approaches, which doesn't contribute

[4]Others is one of the five sectors IPCC uses in its methodology. One example of "others" is indirect emissions from atmospheric depositions.

[5]As pointed by the UNFCCC in its definition: an emission factor is defined as the average emission rate of a given GHG for a given source, relative to units of activity. These units of activities may be, for example, amount of fuel burned for electricity generation or in transport; amount of waste generated per habitant; and hectares of forest areas turned into other types of land (pasture, urban area, etc.).

Table 1 Some tools used by cities to elaborate GHG emission inventories

Tool	Characteristics	Examples of uses
CO_2 Grobillanz/ EMSIG	EMSIG—Emission Simulation in Gemeinden (Emission Simulation in Communities) was developed by Austria's energy agency. CO_2 Grobillanz is a simpler version. They come with data from Austria regarding emission factors, goods consumption, and economic activity-related emissions. Both use geographical frontiers as boundaries	Communities in Austria
ECO_2 region	Supports the calculation of public authority's and/or territory GHG emissions. The framework is mostly compatible with the IPCC (2006) methodology. It is also possible to include emissions of local pollution such as particulate matter. Average emission factors for some countries are included	Cities in Germany, Switzerland, and Italy
GRIP	The Greenhouse Gas Regional Inventory Protocol was developed by the University of Manchester and the United Kingdom's environmental agency. Initially it was designed for metropolitan areas, but it has been used for smaller cities too. The methodology used follows the IPCC Guide—2006, allowing greater comparability between cities. A tool for scenario construction is also available	Cities in the United Kingdom as well as in some other places of Europe and the United States
Bilan carbone ® collectivités—territoires	This tool was developed based on work from the French environmental agency. It supports accountability for all gases included in the Kyoto Protocol, as well as chlorofluorocarbon (CFC) and water vapor emitted by airplanes. French cities' emission factors are available	Municipalities in France
CO_2—Beregner	A result of the work of the Denmark environmental agency in cooperation with a private consulting group, this instrument considers cities as a geographical entity—even though it may be adapted to account solely for the local authority. Only CO_2, CH_4, and N_2O are supported, but the reporting framework follows the IPCC (2006) guidelines. It requires a great range of data, enabling complex inventories. Furthermore, the tool comes with a guide with 37 possible mitigation actions and their impacts may be calculated	Cities in Denmark
Project 2°	This project is a cooperation between the Clinton Climate Initiative, ICLEI, and the Microsoft Corporation. It is based on HEAT, a tool developed by ICLEI. Therefore, the resulting inventories are consistent with IEAP. All six Kyoto Protocol gases are supported and the methodology used is in accordance with the IPCC Guide—2006. One may account emissions for the territory or the governmental authority. Additionally, emission separation in scopes (1, 2, 3) is possible	C40 (a network of the world's megacities committed to addressing climate change)
CACPS	The Clean Air and Climate Protection Software was developed by ICLEI and follows the IEAP model. This software supports the accountability of traditional air pollutants as well as GHG. It also assists in the elaboration of emission reduction strategies through the evaluation of policies and action plans	Mostly by cities in the United States, but also elsewhere

(continued)

Table 1 (continued)

Tool	Characteristics	Examples of uses
GPC	Provides a framework for accounting and reporting city-wide greenhouse gas emissions. The tool was finalized after a pilot test in 2013 and global public comments in 2012 and 2014. It replaces all the previous draft versions of the GPC and supersedes the International Local Government Greenhouse Gas Emissions Analysis Protocol (community section) published by ICLEI in 2009 and the International Standard for Determining Greenhouse Gas Emissions for Cities that was published by the World Bank, United Nations Environment Programme (UNEP), and UN-HABITAT in 2010. Several programs and initiatives have adopted the GPC, including the Compact of Mayors, carbon Climate Registry, CDP, among others	To date, more than 100 cities across the globe have used the GPC (current and previous versions)
Climate action for urban sustainability (CURB)	This is the newest tool from this list and was designed to enable cities in developing and developed nations to identify and prioritize carbon abatement opportunities within their communities. This modeling tool seeks to be able to help municipal governments and local climate planners assess which strategies make the most sense from a cost and performance perspective	Launched in April 2015—designed to be used by cities in developing and developed nations

Source: Based on [1]

for a consistency between one and another. Besides these issues, a number of communities do not provide GHG inventories regularly, hindering comparison over time and action planning for emission reduction [14]. And because of that cities are seen as tools to fulfill the lack existents of an international political framework for climate change, quantifying and monitoring, in a transparent and credible way, the emission reductions by some comprehensive mitigation activities [2].

Mitigation Accounting

Several cities around the world have implemented GHG monitoring systems to track the performance of mitigation actions and policies. For example, Chicago has committed to individual targets for a series of bottom-up mitigation actions, each of which is designed to achieve a specific reduction in GHG emissions. San Francisco, in contrast, has committed to a top-down citywide GHG reduction target. For San Francisco, tracking the GHG effects of individual mitigation actions is not essential to meet its goals, but is useful for informing its actions. New York City has an absolute citywide GHG reduction target, but also tracks the GHG effects of individual mitigation actions to demonstrate progress in annual reports and to assess the effectiveness of the city's policies in reducing emissions. Mexico City

Table 2 Some examples of emission boundaries reported by cities

Cities	Inventories		Scope				Observations
	City emission	Operational emissions	1	2	3	Uncategorized	
Boston	X	X				X	
Denver	X	X				X	Some indirect emissions are included, such as air transport, waste disposal outside the city and emissions embedded in concrete, food and human consumption of water in the inventory of the city's emissions
Dublin	X	X				X	
Minneapolis	X	X				X	Some indirect emissions are included, such as aviation
Hong Kong	X					X	Only direct emissions are accounted for
São Paulo	X					X	Some indirect emissions are accounted for as imported electricity, but no rating scopes. Operational issues were not separate, but it was estimated emissions for the energy consumption of municipal schools and public lighting and the consumption of fossil fuels by the city
London	X					X	
Nova York	X	X	X	X	X		
Portland	X					X	The inventory includes direct and indirect emissions, but without separating them in scopes
Sidney	X	X				X	
Toronto	X	X				X	Some indirect emissions are included, such as the disposal of waste outside the city

Source: Based on [1]

Table 3 Some emission sectors reported by cities

Cities	Sectors
Boston	Residential, commercial/industrial, transport, waste/wastewater
Denver	Buildings and facilities, transport, use and dispose of materials
Dublin	Residential, commercial/industrial, transport, waste
Minneapolis	Residential, commercial/industrial, transportation, solid waste
Hong Kong	Electricity, transport, and other end-use fuels, waste, IPPU, AFOLU
São Paulo	Energy (divided into transport, residential, commercial, industrial, and public), IPPU, AFOLU, solid waste, and effluents
Nova York	Buildings, transport, street lighting and traffic signals, fugitive emissions, and processes
Portland	Energy, transport, materials (goods and services)
Toronto	Residential, commercial/industrial, transport, waste

Source: Based on [1]

has dual goals, tracking progress toward its citywide reduction target as well as generating carbon credits from mitigation projects through the Clean Development Mechanism (CDM).[6]

4 Rio de Janeiro's Experience

Inventory

What is seen as a pioneer action in Brazil and Latin America, Rio prepared in 2000 the first GHG inventory for a city. This study was based on data collected from different years: 1990, 1996, and 1998 [15]. The main methodological challenge was the adaptation of the IPCC Guide to consider emissions resulting exclusively from socioeconomic activity of the city [14, 16]. Another challenge surrounded data collection, which was seen by the researchers as the most delicate stage of the inventory. This involved problems of receiving relevant data from the responsible sectors (local, state, or national) and discrepancies among data obtained from different governmental levels. In order to deal with this matter and overcome these obstacles proxies were used, and also the specialists developing the report were consulted and some strategic choices were made.

In January 2011, Rio enacted the Law no. 5248, establishing the Municipal Policy on Climate Change and Sustainable Development (PMMCDS—in

[6]Despite the new COP 21-Paris Agreement and the end of the Kyoto Protocol and the CDM projects, the new mechanism to be yet fully designed is expected to work on a combination of the Kyoto Protocol/CDM projects and the Kyoto's Joint Implementation program, taking the best of each one and improving over the weakness and moreover aiming for an overall mitigation. Of course those CDM projects ongoing will work as a bridge for this new mechanism and it is expecting better transition between Kyoto and Paris.

Portuguese). Its Article number six states the commitment of reducing GHG emissions by 8 %, 16 %, and 20 % in 2012, 2016, and 2020, respectively—compared to the level of emissions in the city in 2005 [1]. To understand and quantify these goals in terms of volume of greenhouse gases, the city elaborated a new emission inventory for the year 2005 [17] and updated the previous one. With the purpose of comprehending how the economic sectors and actions of the city's government could contribute to this reduction, a study of emission scenarios was also developed for the period of 2005–2030 [17]. In 2013, the city completed the inventory of 2012 and update of 2005 numbers.

In 2011, Rio, in partnership with the World Bank, developed the Rio de Janeiro Low Carbon City Development Program (LCCDP) with the purpose of tracking the performance of policies and actions with a potential to mitigate emissions [18]. This program enables the accounting of reductions accomplished—in order to check the achievement of the targets set in PMMCDS—and also certifies them for a possible commercialization in carbon markets. This program is certified by International Organization of Standardization (ISO) by ISO 14001 and ISO 14064—the ISO certification makes the program a business model that can be replicated in cities all over the world.

The program was initially under the responsibility of the Environmental Secretary; however due to the importance and how it deals with different sectors of administration and strategic management the responsibility was reallocated and the program is now directly under the responsibility of the Mayor's Office [19]. The LCCPD is another example for how the city has been committed to sustainable, low-carbon urban development for the past years.

According to the certification, the program is structured and planned based on two important pillars: Program Roles, and Processes for Program Planning and Evaluation. Each new activity that reduces emissions—called an intervention—goes through the same five-step program process. This procedure ensures the viability of replicating this initiative to other cities in the world and generates transparency and consensus in results to allow for emission trading. To do so it is important to call the attention that depends on the level of engagement of the different municipal institutions. Also the procedures and the teamwork provide a better infrastructure that enables the city to plan, implement, monitor, and become responsible for its own mitigation actions.

Regarding the program roles, Box 2 explains their assignments.

Box 2

The roles of Rio Low Carbon City Development Program—LCCDP [18]

Fixed Assignments.

Coordinating Management Entity (CME): The CME is the central body within the municipality that oversees the coordination and management of the program. Fulfilling this role in Rio is the Mayor's Office (known as "Casa Civil").

(continued)

Box 2 (continued)

Information Management Entity (IME): The IME is the central body that coordinates and manages all information and data related to the program. The IME must ideally have both coordinating capabilities with all municipal departments and experience in collecting and managing large quantities of data. Fulfilling this role in Rio is Instituto Pereira Passos (IPP).

Variable, Intervention-linked Assignments.

Multi-Sector Municipal Working Group (MWG): The MWG is a working group consisting of members from across the municipality with multiple areas of relevant expertise. It acts as an advisory committee to the CME. The composition and attendance of the MWG may vary from intervention to intervention, but it will always be coordinated by the CME.

Technical Advisory Entity (TAE): The TAE is an entity or a consultant with technical expertise in the quantification of emission reductions.

Validation and Verification Entity (VVE): The VVE is an ISO-accredited environmental auditor. It validates and verifies the emission reductions generated by interventions under the program. For any given intervention the TAE and the VVE must not be the same entity to insure integrity in the audit process and avoid conflict of interest.

The program process prescribes the procedures and criteria against which interventions are assessed to be registered in the program, as well as the process of monitoring, reporting, and verifying the emission reductions generated by interventions. The program process steps are presented below (Fig. 2):

The program is still under implementation and is scheduled to contemplate different initiatives until the end of 2016—including green building certification, recycling, reforestation and urban tree planting, bike lanes, and finishing its bus rapid transit system. An intervention under the Low Carbon Development Program can be any activity that reduces emissions, including projects and municipal policies

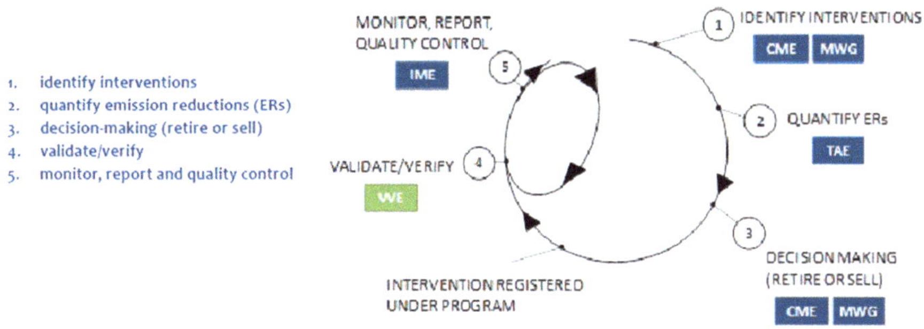

Fig. 2 Five steps of the program process. Reproduced with permission from [18] © World Bank (2012)

in any urban sector. Thus, the program has the potential to extend horizontally and also over time in order to include a wide range of municipal activities [2].

Carloni [19] addresses the successful features of the program:

1. Well-defined framework of roles and processes to plan, implement, monitor, and account for mitigation actions.
2. The organizational structure of the ISO-certified Rio program can be replicated.
3. Flexibility. The program has the potential to expand over time to include a wide range of municipal activities—institutionalizing a "carbon lens" through which ultimately all municipal activities may be viewed.
4. Participation and engagement from municipality staff and stakeholders through a working group of technical experts from the municipality and local university.
5. High-level mandate and engagement under the Office of the Mayor.

And point out that the main constraint is data availability and documentation.

5 Conclusions

The development of GHG inventories is not intended to be an end in itself, but rather a tool for monitoring emissions and a basis for developing strategies to reduce GHG emissions. Quantifying emission reductions of different mitigation actions can be complex and expensive, but following other cities' example and using their strategy as a base point is a good start.

In this sense, Rio de Janeiro's experience can be helpful. The PMMCDS and Rio's Low Carbon City Development Program may be studied and replicated[7] and more information will be aggregated if other cities use this pioneering program.

The city government is investing in guiding their public policies toward a low-carbon urban development. Investments and interventions should have a climate component in its priorities, demonstrating the economic agents and civil society, which is indeed a priority. In addition, the main guideline of the Strategic Plan of the City is to promote sustainable development. The option of City Hall, with the support of the Municipality, was the adoption of realistic and transparent goals of reducing GHG emissions in accordance with the public policy of the city government.

It is important to allow comparability among different studies from different cities and help promote cooperation among them in mitigation and adaptation to climate change. Therefore, it is necessary to bring all actors together, in order to share experiences, and optimize efficiency and tools.

[7]For example, an initiative of the Korean Green Growth Partnership (KGGP), a partnership between Republic of Korea and World Bank Group (WBG).

References

1. Carloni, F.B.B.A.: Gestão do inventário e do monitoramento de emissões de gases de efeito estufa em cidades: o caso do Rio de Janeiro. UFRJ/COPPE, Rio de Janeiro (2012)
2. La Rovere, E.L., Carloni, F.B.B.A.: Mitigando emissões de GEEs no contexto subnacional: o caso da cidade do Rio de Janeiro. In: Fundação Konrad Adenauer (ed.) Cadernos Adenauer XVI - Cidades Resilientes. Rio de Janeiro, Agosto (2015). ISBN: 978-85-7504-193-2
3. Garg, A., Kapshe, M., Shukla, P.R., et al.: Large point source (LPS) emissions from India: regional and sectoral analysis. Atmos. Environ. **36**(2), 213–224 (2002). doi:10.1016/S1352-2310(01)00439-3
4. Peng, C., Wu, X., Liu, G., et al.: Urban air quality and health in China. Urban Studies. **39**(12), 2283–2299 (2002)
5. Ramanathan, R.: Selection of appropriate greenhouse gas mitigation options. Global Environ. Change. **9**(3), 203–210 (1999)
6. Yedla, S., Shrestha, R.M.E., Anandarajah, G.: Environmentally sustainable urban transportation—comparative analysis of local emission mitigation strategies vis-a-vis GHG mitigation strategies. Transp. Policy. **12**(3), 245–254 (2005)
7. Gurney, K.R., Romero-Lankao, P., Seto, K.C., Hutyra, L.R., Duren, R., Kennedy, C., Grimm, N.B., Ehleringer, J.R., Marcotullio, P., Hughes, S., Pincetl, S., Chester, M.V., Runfola, D.M., Feddema, J.J., Sperling, J.: Climate change: track urban emissions on a human scale. Nature. **525**(7568), 179–281 (2015). doi:10.1038/525179a
8. Ometto, J.P.H.B., Bun, R., Jonas, M., Nahorski, Z., Gusti, M.I.: Uncertainties in greenhouse gases inventories—expanding our perspective. Clim. Change. **124**(3), 451–458 (2014). doi:10.1007/s10584-014-1149-5
9. IPCC: IPCC Guidelines for National Greenhouse Gas Inventories. National Greenhouse Gas Inventories Programme/IGES, Japan (2006)
10. ICLEI and the C40. Global Protocol for Community-Scale Greenhouse Gas Emissions (2011)
11. EPA- United States Environmental Protection Agency: Website: Developing a Greenhouse Gas Inventory. https://www.epa.gov/statelocalclimate/developing-state-greenhouse-gas-inventory (2015). Accessed 3 June 2016
12. Dubeux, C.B.S.: Mitigação de Emissões de Gases de Efeito Estufa por Municípios Brasileiros: Metodologias para Elaboração de Inventários Setoriais e Cenários de Emissões como Instrumentos de Planejamento. Tese de D.Sc, Planejamento Energético, COPPE, Universidade Federal do Rio de Janeiro, Rio de Janeiro (2007)
13. Ranganathan, J., Corbier, L., Bhatia, P., et al.: GHG Protocol—A Corporate Accounting and Reporting Standard—Revised Edition. World Resources Institute (WRI) e World Business Council for Sustainable Development (WBCDS). EUA (2004)
14. Neves, C.G., Dopico, Y.B.C.: Análise de Metodologias de Produção de Inventários de Gases de Efeito Estufa de Cidades. UFRJ/Escola Politécnica, Rio de Janeiro
15. Dubeux, C.B.S.E., Rovere, E.L.L.: Local perspectives in the control of greenhouse gas emissions—the case of Rio de Janeiro. Cities. **24**(5), 353–364 (2007)
16. D' Avignon, A., Carloni, F.A., Rovere, E.L.L., Dubeux, C.B.S.: Emission inventory: an urban public policy instrument and benchmark. Energy Policy. **38**(9), 4838–4847 (2010)
17. Rovere, E.L.L., Costa, C., Carloni, F., et al.: Inventário e Cenário de emissões de gases de efeito estufa da Cidade do Rio de Janeiro—2005–2025. COPPE/UFRJ-Secretaria Municipal de Meio Ambiente, Rio de Janeiro (2020)
18. World Bank: The Rio de Janeiro low carbon city development program: a business model for green and climate-friendly growth in cities. Directions in urban development. World Bank, Washington DC (2012)
19. Carloni, F.B.B.A. Low carbon city development program Rio de Janeiro case study. In: Green Growth in Practice: Lessons from Country Experiences Case Studies, Latin America, Brazil. Green Growth Best Practice website. http://www.ggbp.org/case-studies/brazil/low-carbon-city-development-program-rio-de-janeiro (2015). Accessed 3 June 2016

Social Factors Affecting Low-Carbon Cities

Viresh Patel

Abstract Cities are complex anthropological constructs defined by an intricate web of economic, ecological, and social factors. Given their control and influence over planning and services, they play a defining role in climate change mitigation efforts and are well placed to show leadership and foster meaningful transition at various scales. This chapter considers the social factors that affect—both directly and indirectly—and give direction to particular decisions around the strategic planning of, transition to, and ongoing sustainability of low-carbon cities. Three key scenarios are discussed: rapidly emerging cities making early infrastructural decisions, mature cities with high *per capita* emissions, and already built-up cities reconciling growth and emissions in developing countries. The case is made that successful long-term transition toward low-carbon cities requires coherent and effective measures taken across government, industry, and civil society in order to facilitate sustained learning and innovation.

Keywords Low carbon • Civil society • Social • Cities • Community • Emissions • Behavioural change • Infrastructure • Governance

Key Terms

1. **Fuel-poor**. An individual is considered to be fuel-poor when living in a household, living on a lower income, which cannot be kept warm at reasonable cost.
2. **Distributive justice**. Concerns about how the costs and benefits of addressing climate change should be shared.
3. **Procedural justice**. Concerns around making sure that the processes of decision-making are fair, in terms of how, by, and for whom decisions are made.

V. Patel (✉)
School of Geography, University of Oxford, Oxford University Centre for the Environment, South Parks Road, Oxford, OX1 3QY, UK
e-mail: viresh.patel@ouce.ox.ac.uk

S. Dhakal, M. Ruth (eds.), *Creating Low Carbon Cities*,
DOI 10.1007/978-3-319-49730-3_13

1 Introduction

Cities are complex anthropological constructs defined by an intricate web of economic, ecological, and social factors. They play a defining role in climate change mitigation efforts; their control and influence over planning and services place them well to show leadership and foster meaningful transition at various scales. However, the provision of 'hard' infrastructural options alone cannot bring about the large-scale changes required to facilitate low-carbon transitioning, and there is a need to better understand the shaping role of local social contexts.

In the context of low-carbon cities, social factors are those that affect—both directly and indirectly—and give direction to particular decisions around the strategic planning of, transition to, and ongoing sustainability of low-carbon cities. These factors such as ethical values, community responsibilities, and broader cultural institutions are inherently both micro and macro in scale. An understanding of these influences is vital in evaluating the performance and effectiveness of low-carbon cities within specific sociocultural and political contexts around the world.

This chapter reports on the relevance and varying implications of social factors for three key scenarios: *rapidly emerging cities* making early infrastructural decisions; *mature cities* in the developed world with high per capita emissions; and *already built-up cities* in the developing world. Synthesizing key recent literature, it makes the case that a successful long-term transition toward low-carbon cities requires coherent and effective measures taken across government, industry, and civil society in order to facilitate sustained learning and innovation.

2 Rapidly Emerging Cities (*Early Infrastructural Transition*)

An upsurge of urbanisation is driving growth across the developing world, swaying the world's economic balance toward the east and south with considerable scale and speed. In economic terms, a shift in investment away from more established markets to emerging cities and infrastructure can prove difficult to rationalise financially [28]. However, the importance (and associated difficulties) of considering an early shift away from the 'business-as-usual' scenario—and toward the implementation of low-carbon infrastructure—has been acknowledged in recent research surrounding the avoidance of carbon lock-in as a result of technological systems and governing institutions [1].

Most developed countries have recourse to a wide range of instruments in order to manage urbanisation and mitigate increasing energy consumption in cities and their societies. In developing cities and countries, the infrastructural choices we make in the present form foundations for the levels of sustainability that can manifest themselves in the future. In social terms, nascent infrastructural decisions occur and interact at three key levels—household, community, and society-wide institutions among which a diverse range of stakeholders play a role. The social factors influencing each of these levels centre around *who* makes decisions around infrastructure and *for whom* decisions are made. Following Adger et al. [2], it is important to make an early fundamental assertion that sociocultural norms, values, and relationships vary widely within and between societies, alongside having a temporal dimension.

In the context of early rapid urbanisation, Olazabal and Pascual present a valuable discussion about the implications of values and culture of urban society in affecting the prospects of cities in the context of change, making the case for their fundamental role in urban transformation, alongside the flexibility of urban social structure [3]. The authors use Bilbao (Basque Country) as a case study to illustrate the barriers and opportunities of local energy transitioning, extracting key discourses around stakeholders' perceived capacity for change. They build upon the propositions laid out by Adger et al. [2], who posit that the limits to adaptation are contingent upon four social factors: ethics, knowledge, attitudes to risk, and culture. Each of these factors, and their relevance to the scenario of rapidly emerging cities and the early implementation of low-carbon infrastructure, is discussed in the following paragraphs.

Ethical values play a key role in the uptake of low-carbon infrastructure, through directly affecting the governance, social, and economic dynamics of a society. 'Values' in this context refer to intrinsic judgements about what is important and valuable in life, judgements which may not be represented by external material objects. They manifest real-world attributes by way of defining the extent of homogeneity in the preferences of those living in a community [4]. The term *culture* is often applied to a community's shared values. Understanding a society's ethical culture can play a significant role in understanding its limits and barriers to successful infrastructural transitioning. Contextualising any 'business-as-usual' scenario through the social lens of ethics and values offers policy-makers an underlying rationale upon which to better address emission growth. Given the constantly changing nature of sociocultural values, however, difficulty arises in identifying what is regarded as having intrinsic value (see [2]).

Moreover, values become more diverse as scale increases. As we move from household through to community and society levels, an increasing heterogeneity of values causes decision-making to become considerably more complex and potentially contradictory. This lends itself to the idea that initiatives at smaller scales that are accompanied by localised policymaking can play an influential role in both attitudinal and behavioural change, as well as the adoption of new and the associated adoption of new infrastructural options (see Peters et al. [5] for an empirical investigation around the role of local government in engendering community-level action toward a low-carbon strategy). However, it should be clearly recognised that a wealth of other, non-social factors feed into the uptake of new infrastructural options, ranging from familial financial considerations through to a more pragmatic set of practicalities at the household level relating to installation and operational feasibility.

The provisional nature of *scientific knowledge* within society presents potential barriers to early low-carbon infrastructural transition by way of uncertainty and contestation. Controversy around scientific knowledge is associated with the ways in which claims to knowledge about the material world are produced, established, and challenged within scientific communities and subsequently how they enter into wider societal circulation (see also [6]). In their various forms, such knowledge claims inform policy, corporate, and independent decision-making; they impact both directly and indirectly on everyday living. Felt *et al.* discuss the broader social and political issues intrinsically bound up in the domain of scientific risk, opening up provoking concerns raised by innovative technologies, such as the distribution of benefits; the prioritisation of particular scientific areas; the pursuit of particular technological pathways; and the extent of trust by wider society in those actors producing the salient knowledges and making relevant decisions [7]. The subjectivity around scientific decision-making couples with Adger et al.'s notion that different social and organisational cultures, and subcultures, approach foresight of future climate change in different ways [2]. In the context of rapidly emerging cities, these differences in approach toward climate change create difficulties in arriving at a homogenous, society-wide consensus.

The role of knowledge claims is entwined with both ethical values and the perception of risk. Both factors are influenced, in part, by the ways in which scientific knowledge is presented and discussed. Within rapidly emerging cities, the perception of *individual and collective risk* influences decision-making at all levels. Risk in this context refers not only to the perceived threat of climate change, but also to the risks associated with the changes to the status quo in light of any adopted infrastructural transitions at smaller scales, with factors including changes to microeconomic status, the reliability of service provision, and intra- and intercommunity relations. Household- and community-level consumer decisions depend upon the balanced consideration of associated risks. At the household and individual levels, Huijts et al. draw on psychological theories and empirical technology acceptance studies, presenting a framework of energy technology acceptance in which they discuss the importance of perceived risks and benefits, confidence in technologies, and distributive fairness [8]. This micro scale of decision-making constitutes a key consideration in avoiding lock-in status. Moving up in scale to society-wide levels of decision-making, policies around infrastructural transition are also constrained by inertia and cultures of risk denial. This correlates with a distrust and lack of faith in institutions as a contributing social barrier to a fuller public engagement with climate change, as identified by Lorenzoni et al. [15].

Adger et al. suggest that individual actions are shaped by deeply embedded cultural and societal behavioural norms and values [2]. They posit that characteristics at the individual level such as beliefs and preferences, alongside perceptions of risk, knowledge, norms, and values, determine what is perceived to be a limit to adaptation to climate change. This resonates with recent studies of smart grids in developed countries which draw upon insights from social sciences in order to better understand the cultural, social, and psychological factors affecting successful uptake (see [9, 10]). Jackson, discussing social psychological theories of behaviour and change, suggests the gains to be harnessed from identifying with common motivations distinguishable in already established community networks [11]. Moreover, he stresses the inhibiting role of dominant cultural values (alongside institutional barriers) in creating a consumer 'lock-in' to unsustainable lifestyle patterns, before moving on to explore the role of learning in fostering effective changes in behaviour. Darby [12], drawing on evidence from an English village, advances this discussion around social learning as playing a fundamental role in informing consumer choices at household and community levels surrounding particular infrastructural options. Although concerned primarily with domestic energy use, her research demonstrates how energy knowledge is built up over time through a combination of taking action first hand, monitoring usage, and absorbing information from many sources in the surrounding environment (see also [13]). Jackson lends further support to this notion, suggesting that learning through modelling our behaviour on what we see around us and how others behave provides more effective and promising avenues for changing behaviours than information and awareness campaigns [11].

Conversely, a case could be made for a more forceful social approach toward early infrastructural transition (with regard to options being laid out in early urbanisation scenarios) in order to initiate a shift in cultural and societal norms. Therein lies the necessity for an appreciation of the wider social factors at play in order to preclude a 'lock-in' scenario and encourage a thematic cultural shift, thereby nurturing the feasibility of low-carbon infrastructure. Supporting this notion, Madlener and Sunak discuss the directional role of urbanisation in changing consumer needs and behavioural patterns of households [14]. This lends itself to the idea that conscious infrastructural decisions made during early stages of urbanisation can engender shifts *away* from the norm and *toward* innovative, low-carbon infrastructure. At the community level, Peters et al.'s research into the Green Living Centre in London, UK, makes a similar proposition around the initiative's role in establishing new social norms that are more closely aligned with the imperatives of sustainability and low-carbon living [5]. Their paper highlights the enabling role played at the governance and planning levels in creating institutional arrangements that support thematic shifts in consumer behaviour, a finding supported by stakeholder perceptions in a case study conducted by Lorenzoni et al. [15].

3 Mature Cities (*Lifestyle Changes*)

Mature cities are those with already high per capita emissions and conventional infrastructure. In this context macro, society-wide infrastructural changes (both hard and soft) play a fundamental role in any low-carbon strategy. However, this must be paired with an appreciation for micro-level choices made at the household and community levels and, in this respect, many of the elements to be considered in rapidly emerging cities are also applicable to this context.

Whilst the provision of alternative infrastructural and policy frameworks is very much a governance issue, consumer choices will play a pivotal role in the fundamental successes of any such framework. An applied research framework by the Lawrence Berkeley National Laboratory discusses the considerable and often dismissed role of human and social factors in driving technology adoption, policy adoption, and market creation. It suggests that all factors must be understood within a larger cultural context [29]. The authors identify the pivotal role played by micro-level decision-making, in that programmes for energy efficiency services and products tend to be top down rather than bottom up. The problem, they suggest, is that the focus here is based upon extrinsic appeals (such as money or energy saving) rather than intrinsic appeals (such as benefitting the community or making a wider difference), which results in a lack of demand for energy efficiency services. As such, planners and policymakers could benefit from an appeal to both aspects in tandem, which might involve dialogue and collaboration between state and non-state actors in facilitating coordinated operations of mutual benefit (see also [5]). Foxon, drawing on a large interdisciplinary project that in part analyses the social potential and acceptability of transition pathways to a low-carbon electricity future in the UK, outlines a similar point [16]. The research presented attends to closer examination of the ways in which low-carbon governance processes engage with a low-carbon transition, based on an analysis of the tensions and choices faced by key actors (*ibid.*).

In line with this Lorenzoni et al., drawing on mixed-method studies analyzing the perceived barriers to public engagement with climate change in the UK, propose social and institutional change as fundamental factors in energy consumption reduction [15]. They discuss the need to consider individual barriers such as lack of information and lifestyle change, alongside broader institutional arrangements such as lack of enabling initiatives by governments, business, and industry. This supports the proposition developed in the previous section that an appreciation for, and understanding of, stakeholders' values and beliefs is of considerable benefit to those involved in the planning and management of low-carbon transitions. It also bears relevance to the successful uptake of new infrastructural choices at both household and community levels, alongside the formation of policy frameworks conducive to low-carbon transitioning.

Shifting away from these *hard* options involving physical action around energy-efficient infrastructural replacement, mature cities can benefit considerably by creating a fostering environment for lifestyle and behavioural change. In line with calls for an appreciation of the more intrinsic appeals of energy efficiency, for many people the appeal of low-carbon community programmes rests in their provision of wider benefits beyond that related to climate change (see chapter by Matthias Ruth in this volume): for example, improved housing conditions and the opening up of new spaces for social and political innovation [17]. Adopting this perspective, Milner et al. discuss the opportunities within emission reduction programmes for interventions that benefit public health [18]. As noted by the authors, such intrinsic benefits provide an additional rationale to pursue low-carbon strategies at the individual, household, and community levels of decision. Emission reduction initiatives can benefit from an appreciation of these alternative benefits and the underlying motivations that drive behavioural and lifestyle change.

Further to this, in their study of community-based initiatives in the UK Hargreaves *et al.* discuss the role of contextualised knowledge production and a supportive social environment for individual and community decision-making in fostering behavioural change [19]. They suggest the beneficial role of a social environment in which individuals can subject their own routines and behaviours to reflexive scrutiny. There is little indication of larger scale initiatives with behavioural change outcomes, somewhat indicative of associated priority focus at micro-scales of harnessing motivation. At the household level this is supported by various smart metering studies, which make the case for practical learning and a co-evolution of technologies with the social practices and daily routines of users [9].

With climate change being a direct consequence of individual, household, and community behaviour [20], it could be argued that the onus of responsibility rests with these stakeholder groups. However, the means by which this responsibility is distributed within and among communities is of considerable importance. The recent emergence of 'low-carbon community' initiatives has gained increasing attention, particularly in the UK. These initiatives place emphasis on area-based communities as vehicles for low-carbon transitioning, manifesting the idea of shared responsibility and wider implications of individual actions, alongside a visible connection of climate change policy to the everyday practicalities of energy use ([5], cited in [21]). In their 2012 *Viewpoint* publication considering low-carbon communities in the UK, Bulkeley and Fuller discuss how the shift from an emphasis on individual action to community responses has been framed using a discourse around the notion of *justice* [21]. Drawing conclusions from case studies, they outline a disparity between programmes led by government and those led by private and civil society actors, namely that government-led carbon community initiatives seek to manifest benefits at the **fuel-poor** sections of community, whilst civil society schemes place considerably more emphasis on developing community resilience. Building on research by Fraser [22], Bulkeley and Fuller suggest that any policy framework around climate mitigation must take into account marginalised and vulnerable groups; approaches that avoid other issues of social injustice could not be considered 'just'. With respect to procedural justice, the authors problematise the application of the principles of democracy in an arena when expert knowledge claims play a particularly dominant role. Citing Swyngedouw and Heynen [23], the need for 'a more equitable distribution of social power' is proposed.

4 Already Built-Up Cities (*Reconciling Growth and Emissions in Developing Countries*)

Slum urbanisation and informality accompany high rates of urbanisation in the developing world, especially in Africa and Asia. In these contexts, where the social fabric of cities is much more diverse, sustainability transitions through the reconciliation of growth and emission demands a sensitivity to social and political contexts (see [24]). The State of the World's Cities Report acknowledges the rapid growth of urban demographics in developing countries, noting the correlation between urban sustainability and socioeconomic aspects of urban prosperity, namely poverty alleviation, reduced inequality, employment, and investment [25]. This context of growth in developing countries is particularly complex, especially in informal and slum settlements. Here, countries face difficulties in finding ways to negotiate the provision of basic services, such as energy, water, and electricity, in order to navigate toward more sustainable modes of consumption and avoid the conventional reliance upon traditional, inefficient sources of energy.

Peter and Swilling present 'informality' as an issue much broader than slum urbanisation, noting how informality constitutes a range of socio-economic and cultural activities, which extend beyond informal infrastructure into the formal business areas and conventional infrastructure themselves [24]. Rapid growth in developing countries sees people in slums and informal settlements unable to access energy by way of conventional infrastructures. This presents itself as an obstacle when considering the issue of reconciling emissions through infrastructural transition; the provision of these infrastructure has to be coupled with appropriate access. Access to energy infrastructure in developing country contexts is mediated by both the formality of appropriate administrative and governance frameworks at citywide and regional levels, and the informality of existing social structures and informal governance networks. Madlener and Sunak support this notion in their paper on the impacts of urbanisation on urban structures and energy demand, noting how rapidly growing cities in developing countries present a decreasing degree of manageability, and with it a sprawl of illegal housing [14]. Slum settlements often do not allow for the adoption of urban lifestyles or behaviour; their lack of access to a basic infrastructure can result in conditions that are largely rural in lifestyle.

With respect to 'informal' determinants of access to energy infrastructure, take for example the case of India. The country has a strong commitment to renewable energy development and is perceived to be engaging heavily with global climate frameworks in order to promote both rural and urban development in this sector. Yet, research demonstrates the role of existing social structures such as caste and class in defining the terms of access to new energy infrastructure despite their intended community-wide provision [26]. Patel researching in India's westernmost state of Gujarat demonstrates the role of existing and historical sociopolitical power relations in affecting implementation of, and access to, low-carbon energy initiatives. This bears relevance to the work of Bulkeley and Fuller (albeit in a developing country context) surrounding low-carbon communities and social justice, exacerbating the relevance for community-based projects in addressing and appreciating social injustices through low-carbon initiatives. It should be noted that there is dearth of empirical investigation into the social dimensions of low-carbon transitioning in developing contexts (see also [27]). The associated limitations of applying research conclusions from studies in developed countries should therefore be acknowledged and these studies should be built upon as a framework of investigation for these scenarios.

5 Conclusion

Cities play a vital role in climate change mitigation efforts, and are well placed to show leadership and foster change. Their control and influence over planning and services locally can nurture meaningful transition at smaller scales. It is evident that the infrastructural decisions made today will have a much longer term bearing on city resilience into the future.

However, as the review of literature in this chapter has argued, the provision of hard infrastructural options alone cannot bring about the large-scale changes required to facilitate low-carbon transitioning. The individuality of local contextual factors plays a considerable role in defining the relative successes of low-carbon strategies, and there is a need to understand and harness these factors effectively in the planning, transition to, and ongoing sustainability of low-carbon cities. These strategies cannot be temporary, short-term fixes, but instead require coherent and effective measures taken across government, industry, and civil society in order to sustain long-term learning and innovation. The multiplicity of scales and hierarchies across each of these dimensions presents one of the most significant challenges to manifesting any real, long-term transition toward low-carbon cities.

Chapter Summary

- Social factors affect—both directly and indirectly—and give direction to particular decisions around the strategic planning, transition to, and ongoing sustainability of low-carbon cities.
- It is vital to strategise an early shift away from the ‘business-as-usual’ scenario through reconsidering technological systems and governing institutions.
- In social terms, nascent infrastructural decisions occur and interact at three key levels—household, community, and society-wide institutions—among which a diverse range of stakeholders play a role.
- Ethics, the provisional nature of scientific knowledge, perception of risk, and cultural norms are all fundamental social barriers to successful infrastructural transition.
- Actions at the level of governance and planning play an enabling role in creating institutional arrangements that support thematic shifts in consumer behaviour.

- Top-down initiatives tend to focus upon extrinsic appeals (such as money or energy saving) rather than intrinsic appeals (such as benefitting the community or making a wider difference).
- Developing countries face difficulty in finding ways to negotiate the provision of basic energy, water, and electricity services, in order to navigate toward more sustainable modes of consumption and avoid the conventional reliance upon traditional, inefficient sources of energy.
- The multiplicity of scales and hierarchies across government, industry, and civil society presents one of the most significant challenges to manifesting real, long-term transition toward low-carbon cities.

References

1. Unruh, G.C., Carrillo-Hermosilla, J.: Globalizing carbon lock-in. Energy Policy. **34**, 1185–1197 (2004)
2. Adger, W.N., Dessai, S., Goulden, M., et al.: Are there social limits to adaptation to climate change? Clim. Change. **93**, 335–354 (2009)
3. Olazabal M, Pascual U: Identifying social determinants of urban low carbon transitions: the case of energy transition in Bilbao, Basque Country. Basque Centre for Climate Change (BC3 Working Paper Series) (2013)
4. Ostrom, E.: Understanding institutional diversity. Princeton University Press, Princeton (2005)
5. Peters, M., Fudge, S., Sinclair, P.: Mobilising community action towards a low-carbon future: opportunities and challenges for local government in the UK. Energy Policy. **38**, 7596–7603 (2010)
6. Whatmore, S.: Mapping knowledge controversies: science, democracy and the redistribution of expertise. Progr. Hum. Geogr. **33**(5), 587–298 (2009)
7. Felt U, Wynne B, Stirling A et al: Science and governance: taking European Knowledge Society seriously. Working Paper. Report of the Expert Group on Science and Governance to DG Research (2007)
8. Huijts, N.M.A., Molin, E.J.E., Steg, L.: Psychological factors influencing sustainable energy technology acceptance: a review-based comprehensive framework. Renew. Sustain. Energy Rev. **16**, 525–531 (2012)
9. Verbong, G.P.J., Beemsterboer, S., Sengers, F.: Smart grids or smart users? Involving users in developing a low carbon electricity economy. Energy Policy. **52**, 117–125 (2013)
10. Gangale, F., Mengolini, A., Onyeji, I.: Consumer engagement: an insight from smart grid projects in Europe. Energy Policy. **60**, 621–628 (2013)
11. Jackson T: Motivating sustainable consumption—a review of evidence on consumer behaviour and behavioural change. Report to the Sustainable Development Research Network. Centre for Environmental Strategies, University of Surrey (2005)
12. Darby, S.: Social learning and public policy: Lessons from an energy-conscious village. Energy Policy. **34**, 2929–2940 (2006)
13. Darby, S.: The Effectiveness of Feedback on Energy Consumption—A Review for DEFRA of the Literature on Metering, Billing and Direct Displays. University of Oxford, Environmental Change Institute (2006)
14. Madlener, R., Sunak, Y.: Impacts of urbanization on urban structures and energy demand: what can we learn for urban energy planning and urbanization management? Sust. Cities Soc. **1**, 45–53 (2011)

15. Lorenzoni, I., Nicholson-Cole, S., Whitmarsh, L.: Barriers perceived to engaging with climate change among the UK public and their policy implications. Glob. Environ. Chang. **17**, 445–459 (2007)
16. Foxon, T.: Transition pathways for a UK low carbon electricity future. Energy Policy. **52**, 10–24 (2013)
17. Mulugetta, Y., Jackson, T., van der Horst, D.: Carbon reduction at community scale. Energy Policy. **38**, 7541–7545 (2010)
18. Milner, J., Davies, M., Wilkinson, P.: Urban energy, carbon management (low carbon cities) and co-benefits for human health. Curr. Opin. Environ. Sustain. **4**, 398–404 (2012)
19. Hargreaves, T., Nye, M., Burgess, J.: Social experiments in sustainable consumption: an evidence based approach with potential for engaging low income communities. Local Environ. **13**(8), 743–758 (2008)
20. Larsen, K., Gunnarsson-Östling, U., Westholm, E.: Environmental scenarios and local–global level of community engagement: environmental justice, jams, institutions and innovation. Futures. **43**(4), 413–423 (2011)
21. Bulkeley, H., Fuller, S.: Viewpoint—Low Carbon Communities and Social Justice. Joseph Rowntree Foundation, UK (2012)
22. Fraser, N.: Justice Interrupts: Critical Reflections on the 'Postsocialist' Condition. Routledge, New York (1997)
23. Swyngedouw, R., Heynen, N.C.: Urban political ecology, justice and the politics of scale. Antipode. **35**(5), 898–918 (2003)
24. Peter C, Swilling M: Sustainable, low-carbon cities in the 21st century—making it happen! Position Paper for the Summit for the Urban Environmental Accords (2011)
25. SOWCR: State of the World's Cities 2012/13: Bridging the Urban Divide, United Nations Human Settlements Programme. UN HABITAT, London, Sterling, VA (2013)
26. Patel, V.: Just lip service? Insights into the sustainable development prospects of the CDM in rural Gujarat, India. Cardiol. Manage. **4**(3), 291–307 (2013)
27. Chester, M.V., Sperling, J., Stokes, E., et al.: Positioning infrastructure and technologies for low-carbon urbanization. Earth's Future. **2**, 533–547 (2014)

References (Web)

28. Unlocking the potential of emerging-market cities. McKinsey Quarterly. September 2012. http://www.mckinsey.com/insights/winning_in_emerging_markets/unlocking_the_potential_of_emerging-market_cities. Last accessed 4 Nov 2013
29. Toward Building Scalable Low Carbon Cities Through Implementation of a Whole Systems Approach in Several Learning Cities in California—An Applied Research Framework by Lawrence Berkeley National Laboratory in Collaboration with Empowerment Institute for the Cool City Challenge. http://www.empowermentinstitute.net/index.php/societal/cool-city-challenge. Last accessed 8 Nov 2013

Viresh Patel is a D.Phil. Candidate at the School of Geography and the Environment, University of Oxford, where his research lies at the intersection of youth, education, and employment in rural Gujarat, India. He holds an M.A. from the University of Cambridge and an M.Sc. (Distinction) Environmental Policy from the University of Oxford.

Key Drivers and Trends of Urban Greenhouse Gas Emissions

Abel Chavez and Joshua Sperling

Abstract This chapter focuses on the anthropogenic greenhouse gas (GHG) emission trends from urban energy systems and the drivers of urban energy-related GHG emissions. As much as possible, quantitative information on GHG emissions in cities are compiled and key insights are shared on why and how they vary within and across cities. Place-based case studies from global cities are also used to provide relevant context. Although this section is not intended to replace a comprehensive literature review or the wealth of research and information developed and under development, it does aim to gather the current state of knowledge on urban GHG drivers. The synthesis is organized into five sections reflecting current knowledge on urban energy-related GHG emissions: (1) drivers, (2) trends, (3) potential transitions, (4) accounting, and (5) governance. Such background helps to explore three principal questions: (1) What are the drivers and variances for these GHG emissions? (2) What are the sources and types of GHG emissions as a result of urban energy systems and their transitions? (3) What role do various societal actors and governance systems have in shaping transitions in urban energy and GHG emissions?

Keywords GHG drivers • Urban GHG emissions • Energy use

1 Drivers of GHG Emissions from the Energy System

GHG emissions are being produced from urban energy systems that support various engineered infrastructure sectors serving cities. These include transportation, energy generation/distribution, buildings, water supply, waste/wastewater

A. Chavez (✉)
Center for Environment and Sustainability (ENVS), Western State Colorado University, 600 N Adams Street, 81231 Gunnison, CO, USA

Sustainable and Resilient Communities, Western State Colorado University, 600 N Adams Street, 81231 Gunnison, CO, USA
e-mail: achavez@western.edu

J. Sperling
National Renewable Energy Laboratory (NREL), USA

S. Dhakal, M. Ruth (eds.), *Creating Low Carbon Cities*,
DOI 10.1007/978-3-319-49730-3_14

management, industrial production, and agricultural production [1]. Findings from a selection of global, national, and city-level reports and research papers also suggest that the main drivers of GHG emissions from urban energy systems also range from and differ by per capita economic production and consumption growth; changes in GHG intensity of the energy supply; population growth; technological innovation and diffusion; behaviors that can affect choices, lifestyles, and consumption preferences; geography, climate, and resource endowments; and governance approaches (IPCC WGIII Ch. 5, [2–6]).

Though detailed assessments on urban drivers are limited by comparable city-level data, relevant conceptual frameworks for classification of drivers have been proposed. As one example, the Global Energy Assessment (GEA) proposes the following as major types of drivers: natural environment; socioeconomic characteristics; national/international urban function and integration; urban energy systems characteristics; and urban form [4]. Similarly, Romero-Lankao [7] cites Ehrlich and Holdren's IPAT equation developed in the 1970s that includes population, affluence, and technology, and yet also elaborates on this, noting the importance of "... values, lifestyles, policies and other institutional, cultural, ecological and economic determinants that shape urban activities responsible for emissions." In the [5] report, a figure (shown in the Appendix) depicts the interconnections of how GHG emissions, immediate drivers, underlying drivers, policies, and GHG mitigation measures can be assessed in the future, perhaps at different scales. Most recently, Creutzig et al. [3] use a sample of 274 cities to empirically examine the effects on energy use and GHG emissions resulting from economic (e.g., GDP/capita), structural (e.g., density), and geographic (heating + cooling degree days) parameters.

While scientific knowledge and measurement techniques to assess the extent to which these individual and multiple factors drive GHG emissions remain important lines of inquiry, cross-city comparisons and in diverse economic, geographic, and urbanization contexts are required for understanding of urban-to-global GHG emission future scenarios.

Today, cities can be, and often are, broadly classified as OECD (also Annex 1, developed) and non-OECD (also non-Annex 1, developing, industrializing) cities. The city economies and their respective urban areas and metropolitan regions are uniquely and broadly articulated through a number of inherent differences which may be systemic or individual level, leading to distinct urban energy use and GHG emission profiles (described more in Sect. 2). One of the emerging systemic level trends is industrial/economic structure and the dominance of higher carbon-intensive primary and secondary activities located in many non-OECD urban areas (roughly >30 % of economic output, see Table 1). In contrast, institutional maturity and know-how are attributed to emission regulations implemented in OECD economies [8], such that urban areas in OECD economies generally house less carbon-intensive production activities, and economic output is dominated by tertiary activities (often >80 % of economic output, see Table 1). Urban form is another prominent systemic level characteristic with impacts on transportation energy, land use, and building energy use. A common urban form parameter is

Table 1 Key socio-demographics and energy system parameters for selected urban areas and their respective countries

Part A

	City	Population local[b] *[national, annual growth]*[a]	GDP per capita (PPP USD) local[b] *[national, annual growth]*[a]	Population density (people/km^2)	Employment intensity (jobs/cap)	Economic structure[b]
Developed (OECD)	Berlin, Germany	4,301,638 *[81,776,930; −0.05 %]*	$33,311 *[$33,565; 1.02 %]*	3,735	0.491	Primary: 0.3 % Secondary: 18.9 % Tertiary: 80.8 %
	Toronto, Canada	5,936,352 *[34,126,547; 1.04 %]*	$43,905 *[$35,222; 0.82 %]*	2,639	0.504	Primary: 0.4 % Secondary: 18.4 % Tertiary: 81.3 %
	Denver, CO, USA	2,646,782 *[309,349,689; 0.92 %]*	$56,848 *[$42,079; 0.62 %]*	1,536	0.476	Primary: 2.2 % Secondary: 10.6 % Tertiary: 87.3 %
	New York City, NY, USA	19,128,439 *[309,349,689; 0.92 %]*	$63,238 *[$42,079; 0.62 %]*	2,050	0.449	Primary: 0.1 % Secondary: 8.2 % Tertiary: 91.8 %
	San Francisco, CA, USA	4,444,474 *[309,349,689; 0.92 %]*	$68,974 *[$42,079; 0.62 %]*	2,366	0.449	Primary: 0.5 % Secondary: 16.3 % Tertiary: 83.2 %
Industrializing (non-OECD)	Manila, The Philippines	11,845,445 *[93,260,798; 1.88 %]*	12,979 *[$3,560; 2.78 %]*	10,543	0.384	Primary: 0 % Secondary: 28.5 % Tertiary: 71.4 %
	Beijing, China	21,067,850 [1,337,705,000; *0.58 %*]	$20,275 [$6,819; *9.39 %*]	11,516	0.293	Primary: 1.6 % Secondary: 22.6 % Tertiary: 75.7 %
	Delhi, India	22,247,650 *[1,224,614,327; 1.50 %]*	$9,499 [$3,073; *5.79 %*]	11,042	0.287	Primary: 0.8 % Secondary: 14.1 % Tertiary: 85 %
	Bangkok, Thailand	11,190,037 [69,122,234; *0.90 %*]	$23,448 [$7,673; *3.34 %*]	6,436	0.615	Primary: 4.9 % Secondary: 44.6 % Tertiary: 50.5 %
	Bogotá, Colombia	8,868,395 [46,294,841; *1.52 %*]	$15,891 [$8,479; *2.48 %*]	13,514	0.509	Primary: 0.2 % Secondary: 22 % Tertiary: 77.8 %

PART B

	City	Vehicle ownership; vehicles per 1000 people *(annual growth)*[d]	Passenger vehicle use, local (VKT/cap)[e]	Passenger vehicle use, local (VKT/car)[e]	Public transit use, local (km/cap)[e]
Developed (OECD)	Berlin, Germany	572 *(−0.08 %)*	3,071	8,665	1,736
	Toronto, Canada	607 *(0.84 %)*	5,493	11,828	1,050
	Denver, CO USA	797 *(0.02 %)*	11,465	18,209	205
	New York City, NY, USA	797 *(0.02 %)*	8,107	18,260	1,266
	San Francisco, CA, USA	797 *(0.02 %)*	12,722	21,300	810

(continued)

Table 1 (continued)

PART B					
Industrializing (non-OECD)	Manila, The Philippines	30 *(−1.60 %)*	337	16,671	101
	Beijing, China	58 *(16.84 %)*	2,956	14,162	2,781
	Delhi, India	18 *(7.08 %)*	61	2,648	1,872
	Bangkok, Thailand	157 *(1.80 %)*	1,813	10,349	642
	Bogotá, Colombia	71 *(4.26 %)*	612	6,853	3,176

Assembled by authors, from various sources: [a]WRI CAIT; [b]Brookings Institution; [c]Carbon Monitoring for Action (CARMA). http://carma.org; [d]The World Bank Data; [e]The International Association of Public Transport (UITP)

population density, which is observed to be lower among automobile-dependent areas (e.g., US cities) compared to European urban areas, and more so compared to industrializing cities—resulting in significantly lower transportation energy use in non-US cities (131 gallons/capita) vs. US cities (446 gallons/capita) [9] (also see Table 1). Individual-level differences are often witnessed at the household level (e.g., wealth, affluence) or transportation (e.g., vehicle fleet fuel efficiencies), both of which are notably different among the OECD versus non-OECD cities—e.g., New York, Delhi, and Beijing (see Table 1).

The drivers mentioned in this section are particularly nuanced within the context of different energy systems. For example, transportation urban energy systems are directly affected by a number of drivers, including rising incomes, higher vehicle ownership rates, longer trip lengths, and trip frequencies, all of which have implications for vehicular kilometer travel (VKT). Moreover, age of vehicle stocks has a prominent role in transportation energy use and GHGs, with inefficient older vehicles being higher emitters. In urban transportation energy research, for example, per capita income has been a useful determinant of vehicle ownership in different cities globally, as shown by Wang et al. [10], where they note that it is widely accepted that vehicle ownership growth rates rapidly accelerate at incomes around $3000 to $5000 per capita per year (in the year 2000 dollars), and yet differences remain given countries' and cities' unique histories, geographies, and policies. Urban building energy use benchmarks across the USA, the UK, and India also demonstrate the important roles of policy measures (e.g., targeted for existing vs. new buildings) as drivers of building energy use-related GHG emissions [11]. In their study, similarities and differences in energy consumption changes from the 1990s into the twenty-first century and quantitative metrics regarding building energy-use intensities in the three nations are assessed, demonstrating the opportunities and limitations of cross-country and cross-city comparisons for advancing knowledge on drivers.

Perhaps most importantly to note are the differences between the urban energy-related GHG emissions (and their drivers) among developed and rapidly industrializing types of urban areas, and between smaller rapidly urbanizing cities and large well-established megacities. Findings from a number of sources (e.g., WRI, UITP, Brookings) used to assemble Table 1 demonstrate these differences. However, due to minimal resources available for data collection in non-OECD and smaller rapidly urbanizing and industrializing cities, some of these differences, and underlying drivers, remain vaguely understood. While drivers generally relate to the determinants previously described above, more research is needed into urbanization processes and how these drivers of change will shape global GHG emissions. Future work could be optimally informed via a tailor-made typology for these distinct urban areas. Here we aim at highlighting some of the prominent determinants of GHGs found in the current literature, reporting key findings by urban areas within developed and industrializing economies, as illustrated in Table 1. While this summary focuses on a moment in time, understanding the rates of change over time specifically in the context of urbanization, particularly across Asia, Africa, Latin America, and smaller rapidly developing cities everywhere, offers important lines of inquiry for exploring drivers of global GHG emissions from urban energy systems.

2 Profiles and Trends in GHG Emissions from Urban Energy Use

There are two notable trends which can bring understanding to driver of urban GHG emissions: energy use and projected population growth: energy use revealing the demand for services, and population growth exhibiting where new development will occur, thus driving additional demand for materials, energy, and GHGs.

Recent estimates project that global primary energy use reached 525 Quad BTU in 2010, up from 406 Quad BTU in 2000 [12]. Through 2040, world energy use is projected to grow 56 %, to 820 Quad BTU. OECD countries exhibit much of this energy use to date. There are unique trends and opportunities to focus in on global GHG emissions. On the one hand energy use in OECD countries is estimated to increase minimally (17 %), while on the other, non-OECD countries which have had minimal (of total) energy use to date are projected to grow 90 % from 2010 levels. As depicted in Fig. 1, by 2040 non-OECD economies will account for about 70 % of global energy use. In other words, rapidly urbanizing and industrializing non-OECD cities may dominate future global GHG emissions. However, what these energy projections do not consider (of course it would be difficult to do so) are how future city infrastructure development may be altered from a set of baseline projections.

Population growth in urban areas is expected to happen rapidly over the next decades. What is often overlooked though is the placement of the projected urbanization. Ongoing research by Chávez [24] reveals that the bulk of urban population growth through 2030 will occur in small (not mega) cities of less than 300,000 people (see Fig. 2). Understanding how (a) these small(er) cities can champion the next wave of development, and (b) form partnerships in favor of

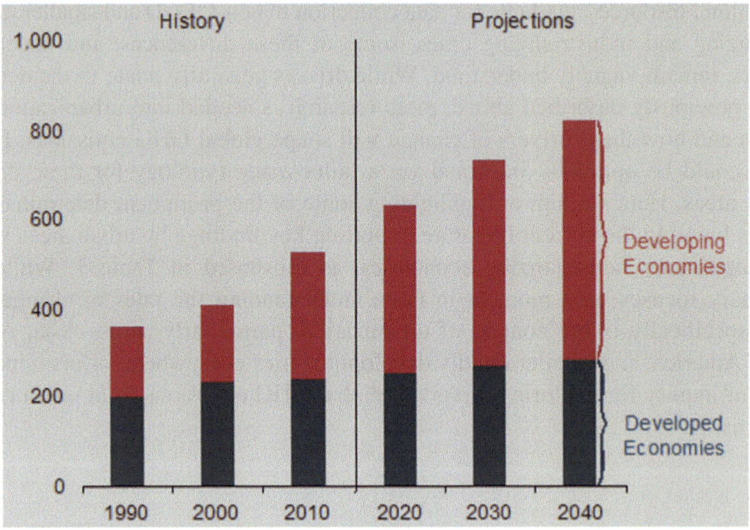

Fig. 1 World energy use from 1990 to 2040 for Developed and Developing Economies. Adapted from (EIA 2013 [12]) International Energy Outlook 2013

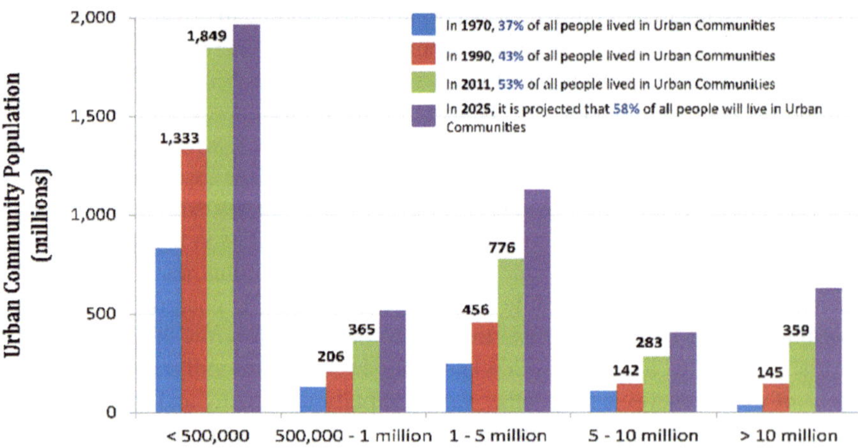

Fig. 2 Urban population by community size. Smaller cities/communities of 500,000 inhabitants or less are expected to house the majority (~50 %) of the world's urban population. Adapted from Chavez (2016)

scaling rapid development, can have substantial implications for the future urban GHG emission drivers.

So where is/will energy use and GHGs occur? Studies have attributed close to 75 % of global GHGs to urban areas based on production-based accounting, and about 80 % of global GHGs based on consumption-based accounting [4, 13];

more on these accounting perspectives is given in Sect. 4. Nonetheless, it is apparent that urban areas and urbanization processes will be core to energy use and GHG trends [25]. Projections summarize that total urban energy-related CO_2 emissions will increase 35 %, to 30 Gt CO_2, between 2006 and 2030 [13]. With rather high rates of growth (e.g., industrialization, urbanization/population, economic expansion), a few key world urban areas in non-OECD (often also classified as developing, emerging, non-Annex 1) economies will shoulder much of this growth.

Assessment of the profiles in urban GHG emissions for specific cities are increasingly being illustrated in the peer-reviewed literature, e.g., for Delhi and Denver [26], or 27 world megacities [14]. As is shown in Table 1, urban GHGs can be influenced by the urban socio-demographic, economic, and energy system characteristics and these trends are of value to the city and global decision makers aiming to address the drivers of urban GHGs.

3 Transition in Technology, Infrastructure, and Fuel Mix in Cities and GHG Implications

As described earlier, as of 2010 the world is mostly urbanized, and in the next 20 years, nearly 60 % of the world's people will be urban dwellers. Even though most developed countries have reached steady urbanization rates, e.g., 80 % (Europe), 75 % (Australia), and 92 % (Japan), trends suggest that urban populations are expected to soar to five billion from more than three billion today and this will include 60 % of China's population living in cities (currently 46 % urban), 41 % of India's population (currently 30 %), and 87 % of the US population (currently 82 % urban) by 2030. Such patterns will lead to shifts in fuel mixes, technological adoptions, and infrastructure requirements, all of which will have GHG emission implications. For example, 615 million people are still without electricity (as of 2011) in developing Asia, including 3 million in China and 306 million people in India (IEA [15], Table 3: Electricity Access in 2011-Developing Asia).

Worldwide, engineered systems are changing and the rate of change may differ significantly in terms of infrastructure, technologies, and fuel mix. In US cities, for example, automobile fleets typically turn over in about one to two decades, and information communication technologies at a much faster rate; buildings remain in the urban stock for many decades [16]; and electricity fuel mixes will also often change based on the ambitions of state renewable portfolio standards that often range from 5 to 20 years (US [17]). Therefore, decisions made on urban infrastructure development and to avoid infrastructure lock-in to fossil fuels as primary inputs will have critical GHG implications.

4 Trans-Boundary Energy Linkages and Their Implications for Urban GHG Accounting

Urban-level GHG emissions are associated with local and global, trans-boundary, supply-chain energy use—extending well beyond the city boundary. Recent literature in the allocation of the urban system boundary has led to the understanding of cities from multiple unique and policy-relevant perspectives, and thus helping explain trans-boundary energy links and GHG implications from these perspectives: Production, expanded-production, and consumption perspectives.

At the core of these perspectives is the "responsibility" of the GHGs associated with a city. A production perspective (also known as "territorial") is in line with the conventional geopolitical definition common to cities, receiving full responsibility for all (industry, commercial, residential, transport) GHG emitted within the settlement. Meanwhile, a consumption perspective attributes all GHGs (local and throughout the supply chain) from local final consumption to the city, whereas any local production that is exported is not allocated to the city, but to the city where the final consumption occurs. A hybrid, expanded territorial, perspective acknowledges that there are additional essential infrastructures serving cities whose GHGs occur outside of the geopolitical boundary, in the hinterlands, and are important to economy-wide development. As such, these distinct definitions reinforce the variety of policy options afforded to cities via their "level of control" or "ownership" per se.

A production perspective subscribes to the fact that urban settlements (and their governments) implement portfolios of policies (e.g., building retrofits) for all buildings, commercial or residential, within their geopolitical boundaries. This is done irrespective of whether an entity produces for local final consumption or exports (external final consumption). On the other hand, policies affecting final consumption afforded to urban settlements range from consumer education campaigns and product labeling to taxes on final products. Implementing successful policies impacting the supply-chain upstream, outside of the urban settlement, may be difficult to achieve for local governments.

In all, one should recall that these frameworks measure different aspects of a "city"; neither is more holistic per se or resulting in a larger GHG estimate, and both are complementary. The differences among these approaches for a particular urban settlement depend on many factors such as economic production technologies, size of export economy, complexity and origins of supply chains, and local final consumption. These differences among perspectives are reflected throughout the literature:

- *Production*: [27], geographic based [28] and is analogous to the *final energy* approach [4] when all final energy is multiplied by the respective GHG emission intensities (or emission factors, EF)
- *Expanded-Production*: [1, 18, 19], Trans-Boundary Infrastructure Supply-Chain Footprint (TBIF) [28], and Community-Wide Infrastructure Footprint (CIF) [26, 27]
- *Consumption*: Carbon Footprint (CF) [20, 27]

5 Stakeholders, Markets, and Governance

Exploring the key actors in carbon management from urban energy systems requires a multidimensional and cross-scale assessment of diverse socio-institutional systems. This includes addressing economic, political, demographic, sociological, behavioral, and historical dimensions of various "who" and "where" questions in the context of emission of greenhouse gas emissions from urban energy systems. Within each of these domains, a simplified framework for assessing key actors, including individuals and businesses, policy actors, as well as infrastructure designers and operators, has been developed [21, 22].

The structure of energy markets in cities also has GHG implications that are often shaped by energy supply, demand, and regulations. Such concepts are often associated with population size, demographic trends, and energy governance. At the city scale, and as demonstrated from various city studies, larger population size is closely associated with higher aggregate urban energy demand and associated greenhouse gas emissions [23]. At different geographic scales, the availability of oil, gas, and coal reserves also shapes geopolitical trends and structures of energy markets for nations and their respective cities.

In order to effectively govern GHG emissions from urban energy systems, multiple levels of government and cross-scale stakeholders play critical roles in the way resources and energy systems are used. Importantly, local governments play important roles in addressing urban energy use and GHG emissions as they can shape zoning regulations, land use, urban form, open space, building codes, modes of transportation, population density, infrastructure design, etc. In addition, cross-scale institutional arrangements, that can include governments at different levels, nongovernment actors, academia, private sector, regulatory bodies, and multiple international institutions (e.g., UN, development agencies, development banks), all have implications for the financial, scientific, legal, and human capital inputs for shifting consumption and production patterns in cities. Perhaps more specifically, the operation of energy infrastructure, electric utilities, and adoption of new forms of energy markets (whether free market, highly regulated, trade restricted, etc.) are all important governance considerations for carbon management from the urban energy system in global cities. Not to be ignored is also the institutional capacity to manage GHG.

Appendix

Figure on GHG drivers [5].

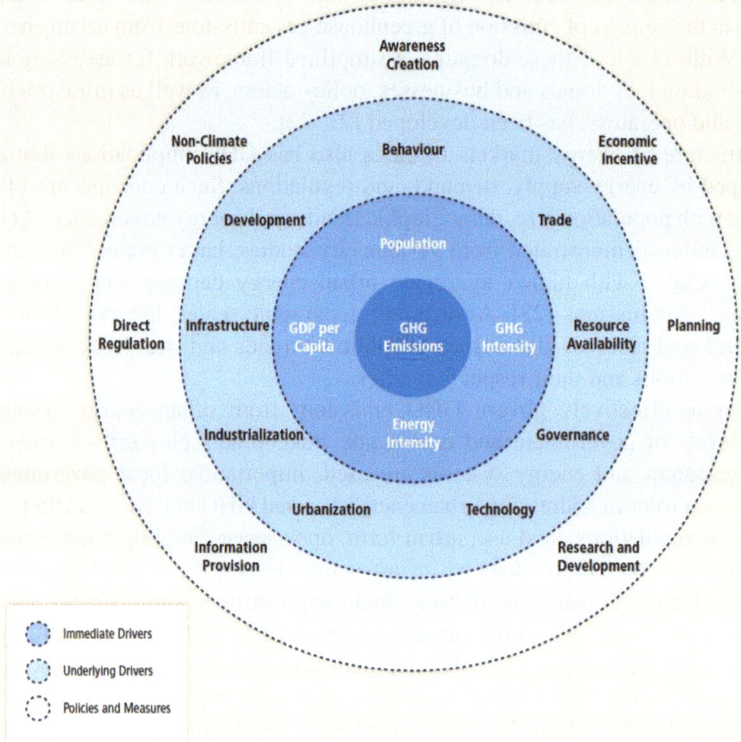

Fig. 3 Interconnections among GHG emissions, immediate drivers, underlying drivers, and policies and measures. Immediate drivers comprise the factors in the decomposition of emissions. Underlying drivers refer to the processes, mechanisms, and characteristics that influence emissions through the factors. Policies and measures affect the underlying drivers that, in turn, may change the factors. Immediate and underlying drivers may, in return, influence policies and measures

References

1. Hillman, T., Ramaswami, A.: Greenhouse gas emission footprints and energy use benchmarks for eight U.S. cities. Environ. Sci. Technol. **44**(6), 1902–1910 (2010)
2. Blodgett, J., Parker, L .: Greenhouse gas emission drivers: population, economic development and growth, and energy use. CRS Report for Congress (2010)

3. Creutzig, F., Baiocchi, G., Bierkandt, R., Pichler, P.-P., Seto, K.C.: Global typology of urban energy use and potentials for an urbanization mitigation wedge. PNAS. **112**(20), 6283–6288 (2015)
4. GEA: Global Energy Assessment—Toward a Sustainable Future. Cambridge University Press; The International Institute for Applied Systems Analysis, Cambridge, UK and New York, NY, USA; Laxenburg, Austria (2012)
5. IPCC.: WGIII AR5, Chapter 5: Drivers, trends, and mitigation (2014)
6. Sattherwaite, D.: The implications of population growth and urbanization for climate change. Environ. Urban. **21**, 545 (2009)
7. Romero-Lankao, P.: Urban areas and climate change: review of current issues and trends. Issues paper prepared for Cities and Climate Change: Global Report on Human Settlements 2011 (2008)
8. Satterthwaite, D.: Sustainable cities or cities that contribute to sustainable development? Urban Stud. **34**(10), 1667–1691 (1997)
9. Newman, P., Kenworthy, J.R.: Cities and Automobile Dependence. Gower, Aldersh (1989)
10. Wang, Y., Teter, J., Sperling, D.: Will China's vehicle population grow even faster than forecasted? Access, Number 41, Fall, 29–33. www.uctc.net/access/41/access41-chinafleet.pdf (2012). Accessed July 2014
11. Gupta and Chandiwala: A critical and comparative evaluation of CO_2 emissions from national building stocks. World Bank, Urban Research Symposium (2009)
12. EIA: International energy outlook 2013. U.S. Energy Information Administration. http://www.eia.gov/outlooks^/ieo/pdf/0484(2013).pdf (2013)
13. IEA: International Energy Agency, World Energy Outlook 2008 (2008)
14. Kennedy, C., Stewart, I., Facchini, A., et al.: Energy and material flows of megacities. PNAS. **112**(19), 5985–5990 (2015)
15. IEA: World energy outlook 2013. International Energy Agency. https://www.iea.org/publications/freepublications/publication/WEO2013.pdf (2013)
16. Reyna, J., Chester, M.: The growth of urban building infrastructure, and its unintended lock-in and embedded environmental effects. J. Ind. Ecol. **19**(4), 524–537 (2015)
17. EIA: http://www.eia.gov/todayinenergy/detail.cfm?id=4850 (2012). Accessed Sept 2014
18. Kennedy, C., Steinberger, J., Gasson, B., Hansen, Y., Hillman, T., Havránek, M., Pataki, D., Phdungsilp, A., Ramaswami, A., Mendez, G.V.: Greenhouse gas emissions from global cities. Environ. Sci. Technol. **43**(19), 7297–7302 (2011)
19. Ramaswami, A., Hillman, T., Janson, B., Reiner, M., Thomas, G.: A demand-centered, hybrid life-cycle methodology for city-scale greenhouse gas inventories. Environ. Sci. Technol. **42**(17), 6455–6461 (2008)
20. Minx, J., Baiocchi, G., Wiedmann, T., Barrett, J., Creutzig, F., Feng, K., Förster, M., Pichler, P.-P., Weisz, H., Hubacek, K.: Carbon footprints of cities and other human settlements in the UK. Environ. Res. Lett. **8**(3), 035039 (2013). doi:10.1088/1748-9326
21. Davis, M., Weible, C.M.: Linking social actors by linking social theories: toward better greenhouse gas mitigation strategies. Carbon Manage. **2**(4), 483–491 (2011)
22. Ramaswami, A., Weible, C.M., Main, D., Heikkila, T., Siddiki, S., Duvall, A., Bernard, M.: A Social-Ecological Infrastructural Systems (SEIS) framework for inter-disciplinary study of sustainable city systems: an integrative curriculum across seven major disciplines. J. Ind. Ecol. **16**(6), 801–813 (2013)
23. Hoornweg, D., Sugar, L., Lorena, C., Gomez, T.: Cities and greenhouse gas emissions: moving forward. Environ. Urban. **23**(1), 207–227 (2011)
24. Chávez, A.: Community Metabolism Footprints: Scaling Communities Towards Global Sustainability (2016)
25. Chester, M., Sperling, J., Stokes, E., Allenby, B., Kockelman, K., Kennedy, C., Baker, L., Keirstead, J., Hendrickson, C.: Positioning infrastructure and technologies for low-carbon urbanization. Am. Geophys. Union's Earth's Future. **2**(10), 533–547 (2014)

26. Chavez, A., Ramaswami, A., Nath, D., Guru, R., Kumar, E.: Implementing trans-boundary infrastructure-based greenhouse gas accounting for Delhi, India. J. Ind. Ecol. **16**(6), 814–828 (2012)
27. Chavez, A., Ramaswami, A.: Articulating a trans-boundary infrastructure supply chain greenhouse gas emission footprint for cities: mathematical relationships and policy relevance. Energy Policy. **54**, 376–384 (2013)
28. Chavez, A., Ramaswami, A.: Progress toward low carbon cities: approaches for transboundary GHG emissions' footprinting. Cardiol. Manag. **2**(4), 471–482 (2011)

Potential Transformation Pathways Towards Low-Carbon Cities

Rose Bailey

Abstract How cities respond to the challenge posed by climate change will fundamentally shape the global response and determine much of humanity's success or failure at dealing with the mitigation and adaptation imperatives. This chapter argues that in order to achieve a desired outcome we need to think differently about how cities plan for long-term transformation to a low-carbon future, and begin by thinking about what we would like to achieve in order to then inform and shape our response. It presents a methodology for generating future scenarios through a hybridised Delphi-backcasting approach with key local actors. It presents two possible scenarios for a low-carbon future and it might be achieved, based on a stakeholder engagement exercise conducted in the city of Bristol, UK. This research sought to gather, refine, and prioritise preferred options for a future city held by core actors in the city, to create shared visions of a successful low-carbon future to inform local decision-making. The chapter then draws some conclusions about the utility of such a method, the results generated, and the implications of participatory scenario exercises for policy and achieving low-carbon futures.

Keywords Scenarios • Backcasting • Low-carbon • Delphi • Bristol • Cities • Futures • Mitigation

1 Introduction

With 54 % of the world's population in 2014 set to grow to 66 % by 2050 [1], 70 % of global emissions are attributed to urban areas, and covering just 2 % of the earth's land area, their footprint far outweighs their physical size [2]. The science of climate change is now unequivocal, and we must significantly reduce global emissions of greenhouse gases by at least 50 % globally and 80 % in the developed world, by 2050, in order to avoid further, considerable disruptions of the climate

R. Bailey MA Oxon. MSc. PhD. (✉)
Ricardo Energy & Environment, Gemini Building, Fermi Avenue, Harwell Oxford, Didcot, Oxfordshire OX11 0QR, UK
e-mail: rose.bailey@ricardo.com

S. Dhakal, M. Ruth (eds.), *Creating Low Carbon Cities*,
DOI 10.1007/978-3-319-49730-3_15

system [3]. What this means for global societies, what the transformation pathway might entail, and what 'low carbon' might look like are highly uncertain, however. This chapter aims to discuss a method by which we might help to define a low-carbon transformation pathway for a city, and provides two examples of transition pathway options and low-carbon futures from the city of Bristol, UK, developed through an innovative stakeholder consultation method. The chapter concludes that the method is useful for consulting and drawing together competing views to generate a shared vision for the future of the city, around which to mobilise key actors and identify the steps needed to achieve a desired future. It finds that participants struggle with the concept of imagining options outside of those currently experienced and accepted, but that even so there is a fundamental disconnect between the future considered 'most likely' and 'most preferable', and the current direction of travel for climate policy by many developed nation states. This therefore provides considerable opportunities for cities to input and support on the behavioural change side, and promote more ambitious options.

2 The Challenge of Climate Change and the Opportunity of Cities

Climate change is unique amongst environmental issues. It presents a complex inter- and intra-generational collective action problem, involving cooperation across multiple scales, actors, sectors, and institutions. Given the relative failure of nation states to tackle the challenges, many cities have been stepping up to the mark and filling the climate leadership vacuum. Proving that sustainable needn't come at the expense of quality of life, a strong economy, and votes, increasing numbers of cities have been acting on the need to reduce emissions [4–7]. The historic Paris Agreement [8], reached at the UNFCCC twenty-first Conference of the Parties (COP21) in Paris in December 2015, marked a turning point for cities as well as national governments. Cities and subnational actors were formally recognised for their role in mitigating climate change.

Despite the efforts of many cities, and their typically lower per capita emissions [9–11], most global cities still have a long way to go to achieve 'low carbon'. To date, and increasingly so recently with the release of new methodologies such as the GPC [12], the focus of much activity has been on either understanding and quantifying the greenhouse gas impact of cities [13]—itself an important first step in managing this impact—or specific areas of intervention such as insulation programmes, or high-profile flagship 'green' projects. Although initiatives such as the 'Compact of Mayors' [14] are going some way now to ensuring that cities undertake a full assessment and action planning process for dealing with both mitigation and adaptation, there are still significant areas for improved understanding in long-term low-carbon transitions for cities [15].

Cities also face additional bureaucratic challenges in responding to climate change. They are complex sites of multiple- and multi-scale actors, agendas, politics, technologies, and infrastructure [16]. Pressures both from above (nationally) and below (communities), fluid and ill-defined boundaries, political short-termism, risk of technological ‘lock-in’, and reliance on larger scale decisions and investments all create challenging circumstances for a city to navigate and balance on a path to a low-carbon future. For example, Erikson and Tempest conclude that new, energy-inefficient urban development may substantially ‘lock in’ future CO_2 emissions, as roughly 30 % of future CO_2 emissions ‘committed’ annually, due to new, urban building and transport systems [17].

‘Low Carbon’: There is no definition of ‘low carbon’, but in policy terms it can relate to a number of different ‘goals’, including the following:

- Base year emission goals—a relative reduction in greenhouse gas emissions (GHGs), often in the order of a 5080 % depending on the development status of the country or city, by 2050 or from some other defined base year.
- Fixed-level goals—an absolute reduction in emissions to a certain level by a target year, such as carbon neutrality.
- Baseline scenario emission goals—usually framed as a percentage reduction in emissions from what might have happened under a business-as-usual scenario.
- Intensity goals—emission reductions per capita or unit of GDP/.

Even defining ‘low carbon’ can be problematic. In the developed world, we often associate ‘low carbon’ with frequently cited targets such as an 80 % reduction in emissions based on 1990 levels (e.g. the UK climate change act [18]), and the political rhetoric claims (and hopes); this is consistent with the IPCC emission scenarios: *‘CO_2-equivalent concentrations in 2100 of about 450 ppm or lower are likely to maintain warming below 2°C over the 21st century relative to pre-industrial levels. These scenarios are characterized by 40 to 70% global anthropogenic GHG emissions reductions by 2050 compared to 2010’* [19].

However in many instances ‘low carbon’ is used simply to mean ‘*lower* carbon’, with cities and countries around the world striving to balance their desire to commit to a low-carbon pathway whist still achieving development objectives. Analysis of pledges made in Intended Nationally Determined Contributions (INDCs) at COP21 for example suggests that whilst the plans represent a significant advance on current trends, which would result in as much as 5 °C of warming if left unchecked, they are not enough in themselves to limit global warming to the 2 °C threshold that countries agreed on [20], showing that there is still a disconnect between what is politically committed to, practically feasible, and scientifically necessary. ‘Low carbon’ also does not automatically result in ‘sustainable’, economically and socially as well as environmentally, and care needs to be taken not to unduly focus on an emission reduction target at the expense of other issues. There exists therefore a balancing act. A strong alignment between low carbon and locally appropriate sustainable development strategies for cities is important, and a low-carbon city must be therefore, above all, a sustainable, efficient, liveable, and competitive city [21].

It should be acknowledged therefore that cities are also all starting from a different baseline and a set of circumstances and objectives, with differing interpretations of goal-setting [22] and working within varying national frameworks for mitigation, and so defining low carbon is unique to each [23].

One problem consistently highlighted, and not only for cities, is the challenge of timescale: identifying long-term transformation pathways towards 'low carbon' [24]. It is not possible to predict the future but it is possible to be confident of certain key trends: growing global population and urbanisation; increased resource pressures and shortages, including energy, water, and food; and continued technological innovation, all over both short and long timescales [25]. There are multiple pathways (see for example the IPCC Special Report on Emission Scenarios [26]), and all it is possible to be certain of is that responding adequately will require significant change.

3 Scenarios for Long-Term Transformation

'Scenario thinking': *'the use of the imagination to consider possible alternative future situations, as they may evolve from the present, with a view to improving immediate and near-term decision making'.*

In order to bring about this significant and meaningful change, stimulate creative and ambitious thinking, and create confidence in actions and outcomes, an alternative perspective is useful. Creating collaborative visions or scenarios and their associated transformation pathways can help to close the gap between 'where we want to be' and 'where we are now', helping to overcome some of the inherent uncertainties by taking a proactive approach to the future. Developed in a participatory way, this process also helps to create implicit buy-in to the results. Scenario approaches are being increasingly used in low-carbon futures research [27–29].

There are many different types of scenarios, serving different purposes [30, 31]. These can be predictive scenarios, involving probabilities and projections of past and current trends, for example population growth or fuel price projections; explorative scenarios, exploring situations or developments that are regarded as possible to happen, such as the Foresight 2020 socio-economic condition scenarios [32], used for strategy planning, Forum for the Future's 2030 climate scenarios [33], or Shell's 2050 energy scenarios [34]; or normative scenarios. The latter are structured around an explicitly normative external constraint (e.g. a target), and are particularly useful for situations involving long-term problems, meeting specific goals, and requiring large-scale transformation and change. They are not about creating a blueprint for the future, but having the inventiveness and reflexivity to create it. 'Transforming-normative' scenarios are particularly helpful in framing responses and developing strategy for long-timescale climate policy [35, 36].

There are a number of city visions or strategies but few meet the definition of 'scenarios', largely being a series of sector-specific targets that themselves form and determine the vision. For example, Vancouver's 'Greenest City 2020 Action Plan' sets goals and targets to achieve by this date [37], and the District of Colombia's 'vision' is also built around a series of sectoral targets [38]. Rio de Janeiro's 'Low Carbon City Development Program (LCCDP)' is a systems approach, including a framework and a set of comprehensive requirements to help the city to plan, implement, monitor, and account for low-carbon investments and climate change mitigation actions across all sectors in the city over time. PlanNYC sets out a 'blueprint' for a sustainable and resilient city but is more a long-term plan, monitoring and reporting system than scenario approach. Other projects such as the World Bank Low-Carbon Liveable Cities (LC^2) Initiative, aiming to help developing country cities progress on a low-carbon development path, focus on specific elements of reaching a low-carbon future, such as finance and emission accounting [39]. Others construct comparative predictive and explorative scenarios based on 'business as usual' and 'low carbon', such as for Bhopal [40]. Forum for the Future's 'Megacities on the move' scenarios offer a set of different explorative scenarios for future mobility in megacities based on different trends and development, particularly looking Mumbai and Istanbul, but lack a specific exogenous constraint to explore futures within [41].

'Backcasting': An approach that starts with defining a desirable future and then works backwards to identify the steps needed to achieve the desired future and connect it to the present.

'Transforming-normative' scenarios offer a combination of these examples, with holistic thinking regarding socio-economic transition, but based around exogenous targets and constraints, and including multiple options for development pathways. They are commonly generated through 'backcasting' approaches [42, 43]. Backcasting involves defining an end point (often within a normative exogenous constraint), which then provides the starting point to work backwards from, identifying the STEPs (social, technological, economic, political interventions) necessary to achieve it (Fig. 1). It is therefore a particularly effective approach for identifying multifaceted low-carbon pathways for cities. Using methods such as Delphi [44]—an iterative opinion-gathering and testing exercise

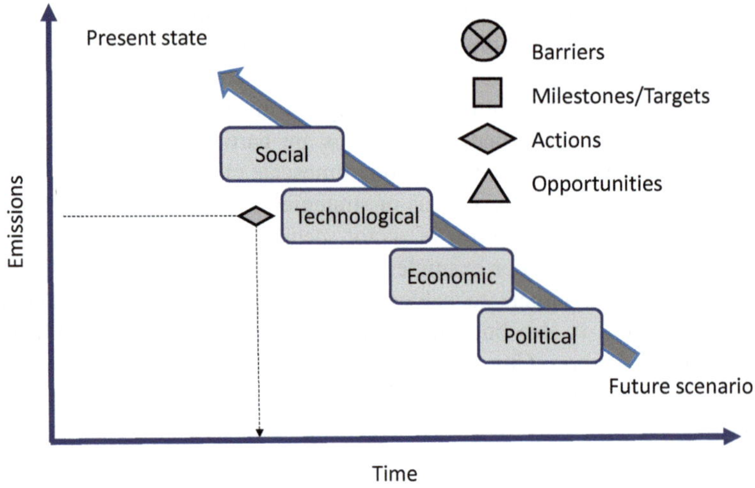

Fig. 1 'Backcasting': starting from a defined future, this method involves mapping out a pathway back to the present

with experts—the process can also help facilitate the combining of views not commonly brought together.

The remainder of this chapter describes the use and outcomes of an innovative combination of Delphi and backcasting methods in the city region of Bristol, South West England, in 2011–2012, to generate possible future low-carbon scenarios and define trajectories to achieve them.

4 Low-Carbon Future Bristol

'Delphi': A method for structuring a group communication process allowing a group of individuals, usually deemed 'experts', to deal with a complex problem. It is a technique for exploring issues using expert judgement, in areas where there are no scientific 'rules'. The experts make some tacit knowledge explicit, and through subsequent rounds the best argument should 'win', and participants should reach a consensus. It is essentially a type of brainstorming used for scenario building, usually conducted remotely, that also has the benefit of avoiding group dynamics.

The research presented here sought to address the issue of long-term city-scale carbon management in the city of Bristol, UK [45]. It specifically focussed on overcoming the difficulties posed by the temporal scale of decarbonisation—

meeting targets some 40 years ahead—through a novel hybridisation of 'Delphi' and 'backcasting' techniques, to identify possible futures consistent with a policy commitment of an 80 % reduction in greenhouse gas emissions from 2005 levels [46]. It explored what a successful low-carbon city region might be like with key local stakeholders, and how such visions might be achieved, and also sought to assess the appropriateness and utility of such a methodological approach in advancing carbon management at the scale of the city.

5 Methodology

This research used Policy Delphi [47] methods with a large stakeholder group to create transformative-normative scenarios for the Bristol region in 2050, within the exogenous constraint of an 80 % reduction in CO_2 emissions. Policy Delphi methods aim to draw out the different arguments rather than a single consensus. They aim to expose all differing positions including the principal pro and contra arguments, estimate the impact and consequences, and estimate the option acceptability [48]. This research intended to achieve these aims, exploring the different options and possible scenarios for the future city, but unlike a true Policy Delphi it still required the generation of consensus, albeit multiple ones.

The selection of participants to create the visions is one of the most difficult and most important aspects when deploying a Delphi method. The use of 'experts' means that the results of a Delphi study are determined by the values they hold, their interests, and biases. The selection of experts to participate in the process was also considered from an output perspective: the process provides an opportunity to make the future rather than just anticipate it, by engaging policymakers and creating implicit support for the results of the process. This can be a useful tool for creating city consensus and commitment across sectors and groups of actors. Participants were identified and selected across key emission sectors, such as energy and transport, and cross-cutting actors such as those working in buildings and spatial planning and climate change policy, and key political and social policy actors (including food, healthcare, and so on). Key interest groups and organisations were also identified. Across these sectors participants were also identified and grouped into categories based on the nature of their expertise: political, managerial/strategic, technical/operational, and research/academic. An example identification and selection process is shown in Fig. 2.

Following a typical Delphi process [49], three rounds of questionnaire were carried out using online questionnaire software:

1. The first questionnaire scoped out the problem and gathered opinions from participants on what the future might look like.
2. From this, a more structured second-round questionnaire was developed, to explore the ideas submitted in the first in more detail, and begin convergence of ideas towards scenarios.

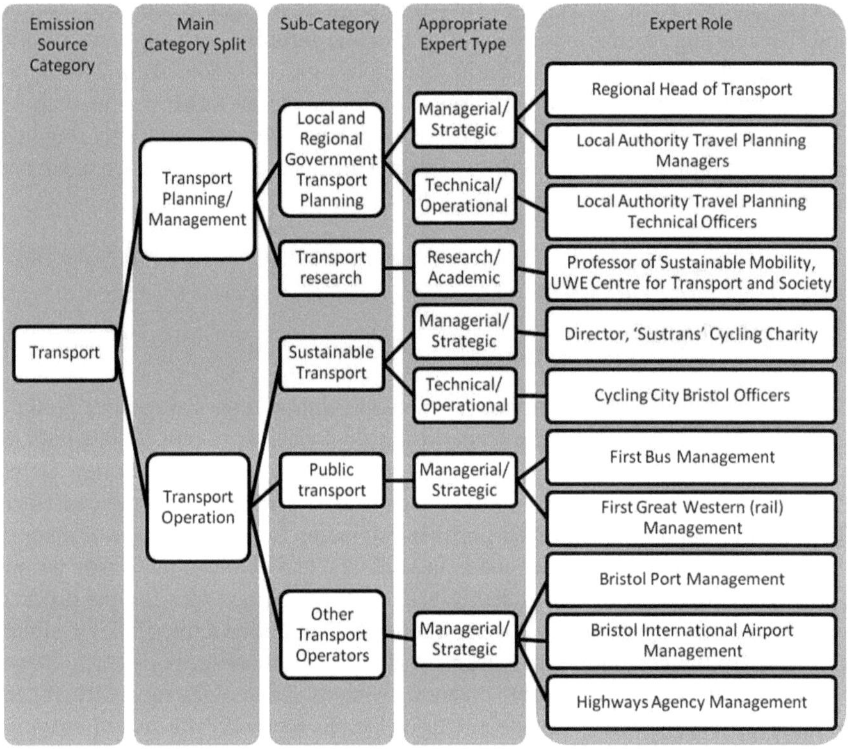

Fig. 2 Example participant selection criteria process

3. The third round of questionnaire presented the provisional scenarios that emerged after the second round, integrating the scenario analysis step of Robinson's (1990) backcasting model [50]. It also acted as a precursor to the final step of determining implementation requirements by investigating key actions and drivers and points in time that would bring them about, for further discussion and analysis in the backcasting workshop.

Open questions from the first questionnaire were analysed using a grounded theory methodology to identify themes, create, and further refine the scenarios. The second and third questionnaires largely used Likert-scale questions to judge preference. These quantitative scores in questionnaires 2 and 3 were statistically analysed using SPSS and cluster analysis [51] to ultimately generate two final scenarios.

From these scenarios, pathways of key actions and drivers that are required to bring the scenarios about were identified with the participants in a backcasting workshop. The backcasting workshop allowed participants to review and validate the scenarios and converge on final agreed sectoral pathways for each (Fig. 3).

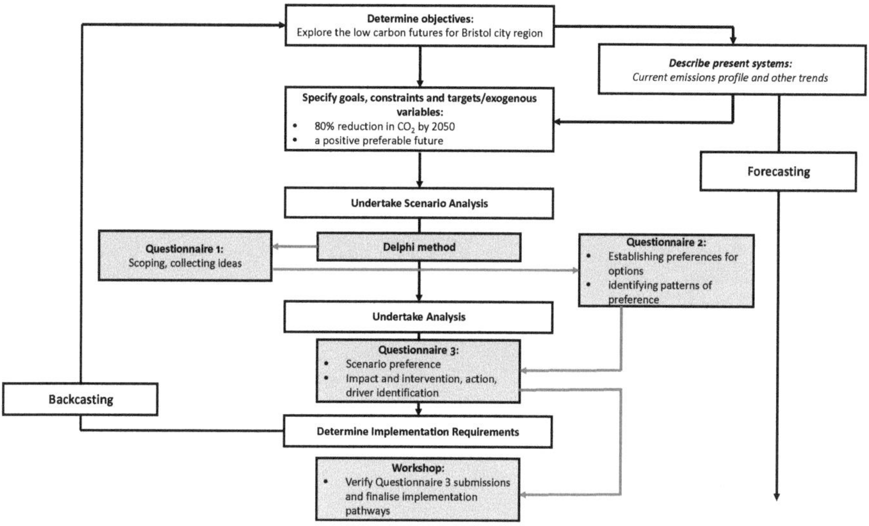

Fig. 3 Research methodology—stages of consultation

6 Outcomes

Two scenarios were developed, with corresponding transition pathways identified to achieve them. Although these scenarios are specific to the Bristol city region, these conceptualisations of contrasting alternative futures—set out below—are broadly transferable and the wider socio-economic and political circumstances present in these different futures may epitomise the contrasting shape of futures that other developed cities may experience and choose to pursue.

The two scenarios for a low-carbon city and their transition pathways can be broadly categorised as 'low-carbon business-as-usual' named 'X', and 're-localised sustainable transformation' named 'Y', and have been visualised on the website www.futurebristol.co.uk (Fig. 4). Although these two scenarios have many differences, there are also many commonalities: re-localisation of services, jobs, and supply chains; high levels of energy efficiency; technological innovation and 'smart' technologies; importance of ICT; high public transport patronage, walking, and cycling; integrated, mixed-use communities; the importance of the high-tech green-tech sector; and sustainability values, prosperity, quality of life, and the role of people and communities. Some key descriptive terms used by the research participants as described above, during the questionnaire process, are shown in the word cloud in Fig. 5, and the key features of the two scenarios are shown in Fig. 6.

Below are the main features of both future scenarios for the city of Bristol, resulting from the stakeholder consultation process.

Fig. 4 The 'Future Bristol' website, depicting two images of alternate low-carbon futures. Reprinted with permission © Andy Council

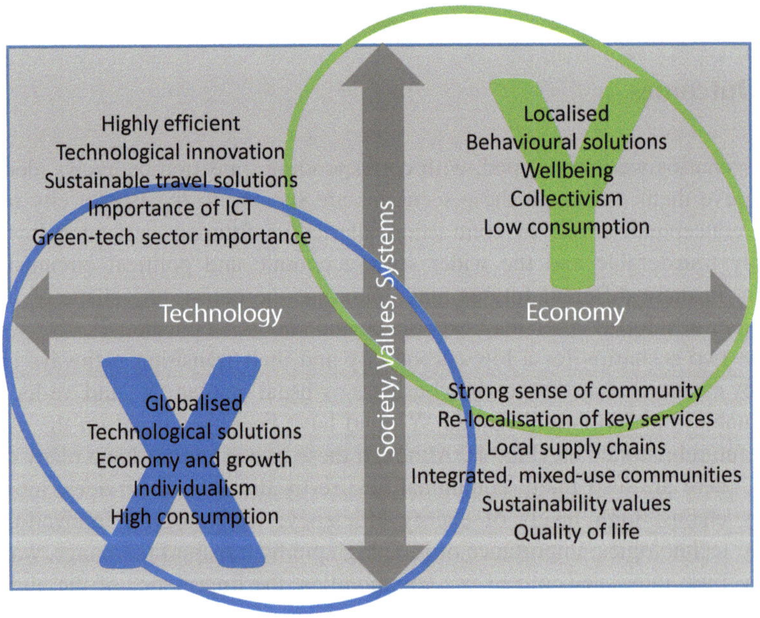

Fig. 5 Word cloud of words most frequently used by the research stakeholder group in the questionnaire responses, in describing their low-carbon future for Bristol

Scenario X: 'Low-Carbon Business-As-Usual'

- Energy: A mix of renewables across the Bristol region provides a top-up to the national decarbonised nuclear and clean fossil fuel base load supply. The region is a centre for the latest in energy efficiency.

Fig. 6 Diagram showing scenario features. In this example, scenarios X and Y sit at the bottom left and top right, respectively. The circles indicate the strength of alignment with the other quadrants, which contain common features of both futures.

- Transport: Travel is still popular, important, and necessary, and a variety of options are available for people in the region, including an efficient and integrated public transport network, and electric vehicles and associated infrastructure.
- Built environment: Traditional style, highly efficient buildings in integrated communities, with some high-rise office hubs, significant green space, and public infrastructure.
- Food, waste, and water: UK produce and some imports, reduced and separated waste, and efficient water use.
- Economy: A thriving, hi-tech economy, internationally competitive.
- Society: A mixed society, environmentally literate, with a good quality of life.

Scenario Y: 'Re-localised Sustainable Transformation'

- Energy: The Bristol region is a leader in decentralised, renewable energy, with most households and communities meeting their needs through integrated generation and high levels of efficiency.
- Transport: The need for travel has been reduced through a move to localisation, but where travel is necessary it is largely by bicycle, foot, or public transport.
- Built environment: Innovative, modern, highly efficient buildings in integrated communities, with significant green space, public transport infrastructure, and urban agriculture.
- Food, waste, and water: Local seasonal produce, reduced and separated waste, respect for water supply.
- Economy: A diverse economy, meeting local needs and providing skills.
- Society: A self-sufficient, collective, slower placed society, with a strong sense of community.

7 Discussion

Two key themes emerged in the analysis of this research.

- Firstly, the apparent difficulty posed by imagining the future, in particular one that is preferable and encompassing the scale of change necessary to address the challenge, including the ability to identify long-term actions. Breaking out of 'probable' thinking to shape an alternative vision is challenging.
- Secondly, the discontinuity between the decarbonisation approaches being pursued at the time and the futures defined as both preferred and perceived as most likely, and the discontinuity between the current policy trajectory and the identified pathways to achieve the scenarios.

The purpose of using a backcasting scenario approach for carbon management was to enable those creating the scenarios to disconnect from the present and think about creating the future they would like to achieve [52]. Policy Delphi approaches allowed this to be realised in a participatory manner, engaging those who are key to bringing such futures about, and therefore hoping to improve legitimacy and the likelihood of success from a policy perspective. The scenarios that were produced, however, cannot be considered either 'radical' or 'transformative', or in the case of X, not even desirable to all, as expected or desired from the process. Instead, the outcomes represent common dichotomies and opinions about the future. Although some comments suggested that elements of the scenarios were 'too extreme', they largely follow existing models and trends, and—at that time—current or already accepted options and technologies. The results imply that participants struggled with the concept of imagining a radically different preferable future outside the boundaries of what is currently considered possible or probable. The challenge of envisaging a preferable future is further evidenced by the presence of some clear trends in responses according to institutional background, suggesting that specific world views and agendas are present within groups [53]. The differing views that were identified between types of respondent raise issues for taking forward and acting on a coherent vision(s): if these key groups have different ideas of what a low-carbon future should look like, they are likely to be working to differing ends and their approaches and priorities may be conflicting. This highlights the importance of methods and outcomes that cross institutional boundaries and challenge world views.

The two scenarios developed through the research represent two possible transformation pathways that a city could choose to take in responding to the need to become 'low carbon'. Despite multiple similarities, they are fundamentally very different, the difference largely residing in the level of proactivity and social transformation required in achieving change. Although the research focussed on the creation of 'preferable' futures, it was clear as the research progressed that there was a disconnect between the 'preferable' and 'probable' or realistic futures imagined by participants. Scenario X was largely considered to be the most 'realistic' scenario, whereas Scenario Y was overwhelmingly the most preferable.

Many respondents were of the opinion that Scenario X was more aligned with national approaches to decarbonisation, and current vested interests and lag times made it the more likely future. It was deemed less preferable, however, as it provides less local opportunity and more state intervention, has a high energy burden, is not considered to be compatible with broader concepts of sustainability, and is heavily reliant on big 'technological fixes'. Scenario Y was considered to be more broadly sustainable, would allow resource shortages such as peak oil to be dealt with better, would bring about a better quality of life, is more resilient and self-sufficient, and has greater local opportunity and 'real localism'. However, it was considered less likely as it requires greater change, from the public, businesses, and governments, and a perceived unwillingness to change limits its feasibility. There is overlap between the scenarios, and neither represents a blueprint for the future. Negative elements of Scenario Y were recognised: some felt that it was a bit too 'extreme' in its localisation, or that the extent of features such as collective ownership of resources would not be as prevalent as the scenario implied, and that greater home working and online services have negative impacts on local economies. Conversely, positive elements of Scenario X were also recognised, such as the importance of technology, the benefits of personal mobility, and the value and continued presence of existing historic buildings.

In generating pathways to achieve the two scenarios, drivers of both were largely considered to occur in the 2020s. From the density of actions identified for the next decade, it was concluded that achieving a preferable local carbon future is critically dependent on actions in the short to medium term. The drivers identified for Scenario Y, the overall 'preferable' scenario, can be summarised into four major areas of short- to mid-term action required to achieve the scenario, all occurring to a large extent throughout the decade from the mid-2010s to the mid-2020s:

1. *The impact of negative external events*: Lack of confidence in nuclear power, energy shortages, high energy prices, and Peak Oil by 2020 drive this scenario, leading to more smaller scale decentralised energy generation by necessity. The failure of large-scale top-down energy schemes (such as new nuclear) was attributed to economic problems, repeat recessions, and the perceived risk of big schemes.
2. *Lifestyle change*: Increasing public awareness of the issue and concerns about limited action were key drivers of this scenario. The need to 'do things differently', make a personal contribution, and live more locally is increasingly recognised by the population over the period to 2050 with collective action becoming more effective than central control in driving change.
3. *Stronger local, integrated planning and leadership*: Stronger planning regulations, better integrated planning, and more holistic planning were all strong features of achieving this scenario. Local government leadership and the maximisation of opportunities available through greater devolved power, and greater community responsibility and spending power, were highlighted as key to bringing this scenario about.

4. *Maximising local economic benefits*: Important in this scenario was identifying and supporting unique local economic sectors and opportunities, in this instance, a large emerging energy retrofit sector and the river Severn Estuary for energy generation.

If the results of this research represent two extremes of a future that might be realised, they may offer a future-proofing capability, particularly if they represent the preferable and the realistic future. The reality, and probably the preferable option for many cities, is likely to be one that merges features of both scenarios. Importantly however, the values and way of life depicted in 're-localised sustainable transformation' are found to be more appealing to people than those of 'low carbon business as usual' and these behavioural factors are one key area where cities and local governments have the ability to shape, support, and encourage, in contrast to their often limited influence over larger supply-side factors.

Discontinuities between the preferable future, and perceived likely future at the city scale, have significant policy implications. Firstly, if local preferences are not compatible with the national decarbonisation agenda, this potentially risks its failure. Public support and buy-in at the local scale are critical to achieve carbon reductions. Although supply-side measures such as new energy generation technologies may deliver many savings, it is well recognised that additional local action will be required, and a strategy that sets out a future that is not compatible with local aspirations is less likely to succeed. The 'additional' climate mitigation benefits of local actions, particularly around supporting behaviour change, should also be recognised, and the success of many of the supply-side measures will depend on local action and support.

Secondly, if Scenario Y is the most preferable future, and one that is judged to be likely to happen, are the UK and other similar developed nations therefore following a pathway with limited chance of success and support? Workshop participants anticipated that a Scenario Y-type future would be the most realistic and achievable because it would happen by necessity. Resource shortages and the anticipated failure of big 'technological fixes' would result in a more localised, resilient, lower consuming future through 'disruptive adaptation' rather than by design. Planning for a future of this type is therefore also a necessity, but in general national climate strategies are relying largely on a few big policies delivering big savings through big technologies. As this research and analysis of the rhetoric versus reality of the climate pledges made at COP21 show, 'closing the gap' between current actions and aspirations becomes even more important, and cities increasingly have a key role to play here.

This sends a clear message that cities must be proactive in trying to achieve a preferred low-carbon future; support the desires of their citizens; challenge conventional orthodoxy, barriers to change, and approaches that are not compatible with local aspirations; and seek power and authority to implement local change. Although this research did not produce a 'radical' vision, a multi-stakeholder consultation exercise such as this can help to foster formation of collaborative visions and trajectories to achieve them, around which cities can catalyse action and shape policy decisions; mobilise key actors, the public, national, and international governments; and be able to demonstrate that the decisions being taken are consistent with a pathway collectively defined that leads towards a desired future. The approach used in this research can also create implicit buy-in to the results through the participation of key actors.

There are ultimately 'multiple pathways' to a sustainable future [54], however that is defined, and the research presented here provides just two examples. It is not possible to predict the future, and all the changes that can and might occur, but by defining a goal first and a desired future state, it is possible to ensure that changes and decisions keep going in the right direction: if possible, towards a preferable future, however that is defined. Scenario and vision planning can help to support this. There is however a fundamental disconnect between the kinds of futures considered 'most likely' and 'most preferable', and the current direction of travel for climate policy by many developed nation states. This therefore provides considerable opportunities for cities, particularly on the behavioural change and advocacy side, to promote more ambitious and preferable mitigation responses.

8 Chapter Summary

- How cities respond to the challenge posed by climate change will fundamentally shape the global response and determine much of humanity's success or failure at dealing with the mitigation and adaptation imperatives.
- There are multiple pathways to 'low carbon', involving differing levels of technological innovation, social change, economic restructuring, and geoengineering, and all we can be certain of is that responding adequately will require significant change.
- Creating collaborative visions or scenarios and their associated transformation pathways can help to close the gap between 'where we want to be' and 'where we are now', helping to overcome some of the uncertainties inherent in the long timescales of carbon reduction by taking a proactive approach to the future. Normative scenarios are particularly useful, and backcasting is a key technique used.
- Although various normative scenarios for a low-carbon future will differ, they often have common elements: re-localisation, energy efficiency, technological innovation, ICT, public transport, walking and cycling, integrated communities, and quality of life for example.
- Two scenarios for a low-carbon city and their transition pathways presented here are 'low carbon business-as-usual', and 're-localised sustainable transformation'. They differ in the level of social change, and the role of large technological fixes.
- Imagining options outside of those currently experienced and accepted was challenging for research participants, but even so there was a fundamental disconnect between the future considered 'most likely' and 'most preferable', and the current direction of travel for climate policy by many developed nation

states. This therefore provides considerable opportunities for cities to input and support on the behavioural change side, and promote more ambitious options.
- Scenarios should not be considered to provide a 'blueprint', but can be a useful tool in engaging key actors and generating a shared vision to catalyse action around.
- The reality, and probably preferable future for many cities, is likely to be one that merges features of both scenarios. Cities must be proactive in trying to achieve a preferred future.

References

1. United Nations: World Urbanization Prospects the 2014 Revision. United Nations, New York (2014)
2. UN-HABITAT: Global Report on Human Settlements 2011—Cities and Climate Change. United Nations, New York (2011)
3. IPCC: Climate change 2014: mitigation of climate change. Contribution of Working Group III to the Fifth Assessment Report of the Intergovernmental Panel on Climate Change. In: Edenhofer, O., Pichs-Madruga, R., Sokona, Y., Farahani, E., Kadner, S., Seyboth, K., Adler, A., Baum, I., Brunner, S., Eickemeier, P., Kriemann, B., Savolainen, J., Schlömer, S., von Stechow, C., Zwickel, T., Minx, J.C. (eds.). Cambridge University Press, Cambridge, UK
4. Betsill, M., Bulkeley, H.: Looking back and thinking ahead: a decade of cities and climate change research. Local Environ. **12**(5), 447–456 (2007)
5. Bulkeley, H., Betsill, M.: Cities and Climate Change: Urban Sustainability and Global Environmental Governance. Routledge, London (2003)
6. http://www.c40.org
7. http://carbonn.org/
8. United National Framework Convention on Climate Change: Report of the conference of the parties on its twenty-first session, held in Paris from 30 November to 13 December 2015. http://unfccc.int/paris_agreement/items/9485.php (2015)
9. Satterthwaite, D.: Cities' contribution to global warming: notes on the allocation of greenhouse gas emissions. Environ. Urban. **20**(2), 539–549 (2008)
10. Dodman, D.: Blaming cities for climate change? An analysis of urban greenhouse gas emissions inventories. Environ. Urban. **21**(1), 185–201 (2009)
11. World Bank: Representative GHG Baselines for Cities and their Respective Countries. http://siteresources.worldbank.org/INTUWM/Resources/GHG_Index_Mar_9_2011.pdf (2011)
12. http://www.ghgprotocol.org/city-accounting
13. Kennedy, C., Steinberger, J., Glasson, B., et al.: Greenhouse gas emissions from global cities. Environ. Sci. Technol. **42**(19), 7297–7302 (2009)
14. https://www.compactofmayors.org/
15. Dhakal, S., Betsill, M.M.: Challenges of urban and regional carbon management and the scientific response. Local Environ. **12**(5), 549 (2007)
16. Bulkeley, H.A., Betsill, M.M.: Revisiting the urban politics of climate change. Environ. Polit. **22**, 136–154 (2013)
17. Erickson, P., Tempest, K.: Keeping cities green: avoiding carbon lock-in due to urban development. Stockholm Environment Institute – U.S. Center. Working Paper No. 2015–11. https://www.sei-international.org/mediamanager/documents/Publications/Climate/SEI-WP-2015-11-C40-Cities-carbon-lock-in.pdf (2015)
18. http://www.legislation.gov.uk/ukpga/2008/27/contents

19. IPCC: Climate Change 2014: Synthesis Report. Contribution of Working Groups I, II and III to the Fifth Assessment Report of the Intergovernmental Panel on Climate Change. In: Core Writing Team, Pachauri, R.K., Meyer, L.A. (eds.) (2014)
20. UNFCCC: Conference of the Parties Twenty-first session Paris, 30 November to 11 December 2015. Synthesis report on the aggregate effect of the intended nationally determined contributions. Note by the secretariat. http://unfccc.int/resource/docs/2015/cop21/eng/07.pdf (2015)
21. Baeumler, A., Ijjasz-Vasquez, E., Mehndiratta, S. (eds.): Sustainable low-carbon city development in China. The World Bank. http://siteresources.worldbank.org/EXTNEWSCHINESE/Resources/3196537-1202098669693/4635541-1335945747603/low_carbon_city_full_en.pdf (2012)
22. Levin, K., Finnegan, J., Rich, D., Bhatia, P.: Mitigation goal standard. An accounting and reporting standard for national and subnational greenhouse gas reduction goals. World Resources Institute. http://www.wri.org/publication/mitigation-goal-standard (2014)
23. Baeumler, A., Ijjasz-Vasquez, E., Mehndiratta, S. (eds.): Sustainable low-carbon city development in China. The World Bank. http://siteresources.worldbank.org/EXTNEWSCHINESE/Resources/3196537-1202098669693/4635541-1335945747603/low_carbon_city_full_en.pdf (2012)
24. Dhakal, S., Shrestha, R.M.: Bridging the research gaps for carbon emissions and their management in cities. Energy Policy. **38**(9), 4753–4755 (2010)
25. Ratcliffe, J., Krawczyk, E.: Imagineering city futures: the use of prospective through scenarios in urban planning. Futures. **43**(7), 642–653 (2011)
26. http://www.ipcc.ch/ipccreports/sres/emission/index.php?idp=0
27. Hughes, N., Mers, J., Strachan, N.: Review and analysis of UK and international low carbon energy scenarios. A joint working paper of the UKERC and the EON.UK/EPSRC Transition Pathways Project. http://www.lowcarbonpathways.org.uk/lowcarbon/publications/index.html (2009)
28. Hughes, N., Strachan, N.: Methodological review of UK and international low carbon scenarios. Energy Policy. **38**(10), 6056–6065 (2010)
29. Bailey, R., Longhurst, J.W.S., Hayes, E.T., Hudson, L., Ragnarsdottir, K.V., Thumim, J.: Exploring a city's potential low carbon futures using Delphi methods: some preliminary findings. J. Environ. Plan. Manage. **55**(8), (2012)
30. Börjeson, L., Höjer, M., Dreborg, K., Ekvall, T., Finnveden, G.: Scenario types and techniques: towards a user's guide. Futures. **38**(7), 723–739 (2006)
31. EEA: BLOSSOM – bridging long-term scenario and strategy analysis: organisation and methods. European Environment Agency. Technical Report no. 5/2011. Office for Official Publications of the European Communities, Luxembourg (2011)
32. DTI: Foresight Futures 2020: Scenarios and User Guidance. http://www.efst.hr/~nalf/Scenario%20analysis%20UK%202020.pdf (2002)
33. Forum for the Future and HP Labs Climate Futures: Responses to Climate Change in 2030. Forum for the Future, London (2008)
34. http://www.shell.com/global/future-energy/scenarios.html
35. Robinson, J., Burch, S., Talwar, S., O'Shea, M., Walsh, M.: Envisioning sustainability: recent progress in the use of participatory backcasting approaches for sustainability research. Technol. Forecast. Soc. Change. **78**(5), 756–768 (2011)
36. Mander, S.L., Bows, A., Anderson, K.L., Shackley, S., Agnolucci, P., Ekins, P.: The Tyndall decarbonisation scenarios—part I: development of a backcasting methodology with stakeholder participation. Energy Policy. **36**(10), 3754–3763 (2008)
37. City of Vancouver: Greenest City 2020 Action Plan. http://vancouver.ca/green-vancouver/greenest-city-2020-action-plan.aspx (2012)
38. District of Colombia: A vision for a sustainable DC. http://sustainable.dc.gov/publication/mayors-vision (2012)
39. http://www.worldbank.org/en/news/press-release/2013/09/25/world-bank-initiative-planning-finance-challenges-300-low-carbon-livable-cities

40. Kapshe, M.: Developing low carbon vision for Indian cities: a case of Bhopal, India. http://www.qub.ac.uk/research-centres/RespondingtoClimateChangeUK-IndianPerspectives/FileStore/Filetoupload,310534,en.pdf (2012)
41. Gazibara, I., Goodman, J., Madden, P.: Megacities on the move your guide to the future of sustainable urban mobility in 2040. Forum for the Future. http://www.forumforthefuture.org/sites/default/files/project/downloads/megacitiesfullreport.pdf (2010)
42. Robinson, J.B.: Unlearning and backcasting: rethinking some of the questions we ask about the future. Technol. Forecast. Soc. Change. **33**(4), 325–338 (1988)
43. Robinson, J.B.: Futures under glass: a recipe for people who hate to predict. Futures. **22**(8), 820–842 (1990)
44. Linstone, H.A., Turoff, M. (eds.): The Delphi Method: Techniques and Applications. New Jersey Institute of Technology, Newark (2002)http://is.njit.edu/pubs/delphibook/
45. Bailey, R.: An exploration of the low carbon futures for the Bristol region. Ph.D., University of the West of England. http://eprints.uwe.ac.uk/17003 (2012)
46. Bailey, R., Longhurst, J.W.S., Hayes, E.T., Hudson, L., Thumim, J., Ragnarsdottir, K.V.: Exploring a city's potential low carbon futures using Delphi methods: some preliminary findings. J. Environ. Plan. Manage. **55**(8), 1022–1046 (2012)
47. de Loe, R.C.: Exploring complex policy questions using the policy Delphi: a multi-round, interactive survey method. Appl. Geogr. **15**(1), 53–68 (1995)
48. Turoff, M. The Policy Delphi. In: Linstone, H.A. and Turoff, M., (eds.) The Delphi Method: Techniques and Applications. Newark: New Jersey Institute of Technology, pp. 80–96. (2002). http://is.njit.edu/pubs/delphibook/
49. Linstone, H.A., Turoff, M.: The Delphi Method: Techniques and Applications. New Jersey Institute of Technology, Newark (2002)http://is.njit.edu/pubs/delphibook/
50. Robinson, J.B.: Futures under glass: a recipe for people who hate to predict. Futures. **22**(8), 820–842 (1990)
51. Tapio, P.: Disaggregative Policy Delphi. Using cluster analysis as a tool for systematic scenario formation. Technol. Forecast. Soc. Change. **70**, 83–101 (2002)
52. Hughes, N.: A historical overview of strategic scenario planning. A joint working paper of the UKERC and the EON.UK.EPSRC Transition Pathways Project. http://www.lowcarbonpathways.org.uk/lowcarbon/publications/Theme1_workingpapers.html (2009)
53. Bailey, R., Longhurst, J.W.S., Hayes, E.T., Hudson, L., Ragnarsdottir, K.V., Thumim, J. Exploring a city's potential low carbon futures using Delphi methods: some preliminary findings. J. Environ. Plan. Manage. 55, 2012 8 (2012)
54. Guy, S., Marvin, S.: Understanding sustainable cities: competing urban futures. Eur. Urban. Reg. Stud. **6**(3), 268–275 (1999)

Rose Bailey is a Principal Consultant specialising in greenhouse gas emissions inventories and management, at Ricardo Energy & Environment, and a former Research Fellow in Carbon Management in the Department of Geography and Environmental Management at the University of the West of England, Bristol. Her specialisms are city and national emission inventories, low-carbon cities, city-scale GHG management, low-carbon futures and scenario studies, and climate change public engagement.

Eco-Districts as a Transition Pathway to Low-Carbon Cities

Jennifer Lenhart and Joan Fitzgerald

Abstract Climate change requires the urgent adoption of low-carbon practices to mitigate greenhouse gas emissions. Nowhere is this more relevant than in cities, which hold over half of the world's population and produce 70 % of all GHGs. This chapter examines whether eco-districts, a growing urban development phenomenon, can serve as a transition pathway to enable low-carbon practices. Using the case of Malmö, Sweden, we assess what role eco-districts can play to enable cities to achieve their climate goals, including whether the lessons of eco-district development are applied to other parts of the city. We also observe how planners and elected officials in Malmö enacted a deliberative process of organisational learning when implementing their eco-district, namely their openness to experimentation with new technologies and planning approaches. We identify how double-loop learning served as a mechanism to support Malmö's eco-district development, in particular when addressing unforeseen barriers to new planning practices. This chapter is based on "Eco-districts: Can they accelerate urban climate planning", published in Environment and Planning C in December 2015.

Keywords Malmö • Eco-districts • Urban climate change planning • Urban climate change governance • Organisational learning • Double-loop learning • Policy diffusion

---

J. Lenhart (✉)
Environmental Policy Group, Wageningen University, Wageningen, The Netherlands

Ulriksdals Slott, WWF Sweden, 10781 Solna, Sweden
e-mail: jennifer.lenhart@wwf.se; jenn.lenhart@gmail.com

J. Fitzgerald
School of Public Policy and Urban Affairs, Northeastern University, 360 Huntington Ave., 310 Ren Park, Boston, MA 02115, USA
e-mail: jo.fitzgerald@neu.edu

S. Dhakal, M. Ruth (eds.), *Creating Low Carbon Cities*,
DOI 10.1007/978-3-319-49730-3_16

1 Introduction

With roughly 70 % of all greenhouse gas (GHG) emissions originating in urban areas, a city's policies and planning are crucial in tackling global GHG emissions [1]. One way to support low-emission development is the creation of an eco-district, or a neighbourhood-scale development that adopts sustainable planning strategies and state-of-the-art technologies in green building, smart infrastructure and renewable energy. Eco-districts can set higher standards for sustainable development throughout a city. As the eco-districts concept diffuses as a prominent sustainability strategy, urban decision-makers need to understand what role they can play in helping cities achieve ambitious climate goals.

To address the eco-districts trend, we ask two questions. First, are eco-districts used by city governments as test beds or learning labs, with lessons learned, applied to other parts of a city, in particular when cities incorporate a deliberate process of organisational learning, allowing city planners and elected officials to apply the lessons learned? If so, then eco-districts have broad potential to accelerate climate change planning. Second, in response to criticisms that eco-districts are merely islands of green privilege, we ask whether the lessons learned can be applied in lower income neighbourhoods.

This chapter builds on existing scientific literature on urban climate change planning, focusing on eco-districts [2–6] and their potential to support a city's long-term climate change planning agenda. Following Sussman [6], we examine how eco-districts can support similar goals in other developments while investigating the importance of learning and experimentation when incorporating urban climate change planning [7–11].

2 Urban Climate Change Planning: A Focus on Eco-Districts

A broad literature has emerged on urban climate change planning, notably what constitutes successful practice [7, 13–18]. This literature discusses the "what" of climate and sustainability planning. We focus on "how", identifying factors accounting for cities' actual GHG emission reductions, by examining governance and planning as important leverage points for transformative change in achieving these reductions [8].

Recently, eco-districts have received attention as an effective way for cities to advance their climate change plans [3, 4]. While eco-districts emerged in the early 1990s, only recently has analysis of their impact appeared in the academic literature. There are two types of eco-districts: those developed on brownfield sites in former industrial areas, and those integrated within existing built areas. Three of the most celebrated eco-districts include *Hammarby Sjöstad* in Stockholm, *Western Harbour* in Malmö and *Vauban* in Freiburg. All were built on brownfields, which, given their scale, had greater potential for experimentation. Each project installed district-scale infrastructure and technology, such as district heating and cooling provided with renewable energy or green storm-water infrastructure, recognised as important to address climate change [16].

The nascent literature analysing eco-districts suggests that their publicity is not supported by actual outcomes. Referring to *Hammarby*, Rutherford [19] suggests that there is more environmental discourse than actual performance. Sussman [6] evaluates Vancouver's *Southeast False Creek* eco-district, stating that it has made limited progress on GHG reduction. Another criticism addresses environmental gentrification, or "sustainability" that drives out low-income residents [20, 21]. Ecological gentrification includes expanding exclusively middle-class housing into previously undeveloped or post-industrial areas [22]. In Stockholm, Rutherford [19] states that a conservative shift in government reduced the makeup of private versus public housing in *Hammarby*, suggesting that it became a green middle-class enclave. In Malmö, critics argue that *Western Harbour* transformed a post-industrial landscape to meet economic goals at the cost of social gains [2, 5].

While the literature contains legitimate critiques, there is little discussion of the longer term potential of eco-districts to facilitate learning or roll out similar goals in a city. Understanding the policy that has led to success or failure is an important, but neglected, aspect of eco-districts.

Organisational Learning in Urban Climate Change Planning and Policy

Effective implementation of a city's climate change plans requires a deliberate process of experimentation, re-evaluation and adaptation of practices [7, 8, 11–13]. Argyris and Schön [10] refer to this process as double-loop learning, or an ongoing modification of organisational values, policies and norms to incorporate new practices. This concept builds on Argyris and Schön's theory of reflexive action in organisational learning [23, 24], in which professionals reflectively integrate their formal knowledge and skills along with other knowledge sources to solve problems. Dieleman [25] argues that reflective action is particularly relevant in networks, where professionals with different backgrounds participate to "realize that creating networks is not just a matter of blending people, technology and knowledge together, but involves learning processes that will often take long periods of time". Dielemen uses Kolb's experiential learning cycle (1984) to develop an education methodology and training package to help cities achieve "eco-cultural innovation". Kolb's cycle [26] includes active experimentation, concrete experiences, reflective observation and abstract conceptualisation, describing learning in these spaces of experimentation.

Eco-districts, because they integrate professionals from different departments of city government and the private sector, offer an ideal planning approach to examine how double-loop learning and the process of reflective practice occur and act to accelerate climate change planning to achieve increasingly ambitious goals. If planning and policy learning are deliberately built into implementation, new technical knowledge can be applied to other city districts and new understanding of how to better incorporate equity—the frequently missing leg of the stool—into other eco-districts can be applied.

3 Malmö's Approach to Climate Change and Sustainable Development

To investigate the eco-district trend described in Sect. 2, several of Europe's most widely celebrated eco-districts were examined as part of a larger study[1] analysing their importance for climate change planning, including Malmö's Western Harbour; Stockholm's Hammarby Sjöstad and Royal Seaport; and Freiburg's Vauban. This chapter describes Malmö's Western Harbour in further detail. When conducting multiple case studies exploring the same questions, one city may emerge as a leader, or critical case [27]. Following a review of policy documents, site visits, interviews, as well as supporting information (e.g. awards[2] and media coverage) Malmö was identified as such a critical case. This chapter uses a case study methodology to examine planning in its real-life context [28]. To effectively analyse the development and continued evolution of Malmö's Western Harbour, several site visits and interview series were conducted over a course of 7 years.

In 2009, data were collected from archival sources, planning documents, and organisational websites on Malmö's Western Harbour to better understand the eco-district and prepare for site visits and interviews. This followed with an extensive tour of the Western Harbour. Interviews were conducted with city planners, elected officials, and practitioners involved in planning and implementation. In 2010, a second visit was conducted, with another round of interviews with three of the five planners initially interviewed. In 2012, a third round of interviews was completed by telephone to ensure that all departments involved were represented. Interview questions focused on the specific strategies employed, how they fit into the city's sustainability/climate action plan, levels of experimentation, and mechanisms employed to learn from successes and failures of practice.

The national background has also influenced Malmö's approach and ability to act as a local climate change leader. Sweden has ranked as one of the world's

[1]"Eco-districts: Can they accelerate urban climate planning", published in Environment and Planning C in December 2015.

[2]Malmö's awards include a 2009 UN-Habitat Scroll of Honour award for sustainable development; a 2010 award from the Building Exchange for the Best Master Plan for the Western Harbour; and a 2010 United Nations World Habitat Award for the revitalisation of Augustenborg. Malmö was the inaugural winner of WWF's Earth Hour City Challenge in 2011. In 2012 Malmö was a finalist for the European Green Capital.

leading countries in climate policy and performance [29][3]. The Swedish EPA has provided substantial subsidies to support local governments to invest in building efficiency, infrastructure and energy systems over the years [30]. Swedish cities also benefit from extensive public land ownership, and local income tax collection, as most social services are locally provided [9]. Combined with the *Local Government Act* in 1991, which grants Swedish cities autonomy to create organisational structures best suited to their duties, cities have greater latitude to achieve sustainability goals, than many of their counterpart local governments in other countries, such as the United States. These vertical integration factors lay the groundwork for success; however local factors related to urban planning practices, as will be described below, play an even larger role.

Malmö (population 300,000) is Sweden's third largest city, and has historically been an industrial centre. Its transformation from industrial city to sustainability city is quite considerable, given the loss of a third of its jobs in the 1980s/1990s. It was with the 1994 election of Social Democrat Ilmar Reepalu as mayor (1994–2013) that Malmö's transition to a knowledge-based sustainable city began. Shortly after his election, Reepalu began a process to engage the city council, businesses and residents in creating a long-term, three-pronged vision to redirect the city. The first prong was to build a knowledge economy, realised with the opening of a new university in 1998. Malmö University now enrols 24,000 students. The second prong was connectivity. Roughly 25 km from Copenhagen, the Oresunds Bridge was completed in 2000, connecting these two cities in a transnational economic region. The third prong was to refocus Malmö as an environmentally sustainable city, to redevelop brownfields as eco-districts. With Western Harbour's development, planning practitioners were encouraged to develop holistic approaches to address all aspects of sustainability simultaneously. Climate change (mitigation and adaptation) planning was implicit in these efforts.

Malmö integrates climate planning into its broader sustainability strategies, notably its *Environmental Programme*, its *Energy Strategy* and its *Master Plan;* however it has not adopted a specific climate policy. These strategies have complementary goals and a pathway for Malmö to become "climate-neutral by 2020", focusing on energy efficiency, renewable energy and integrating sustainability within all municipal activities and infrastructure. The Environmental Programme states, by 2030 "the entire municipality aims to run on 100 percent renewable energy".

[3]Sweden ranked highest in 2008 and 2009 and second highest in 2010, 2011 and 2013.

The Eco-District: From Industrial Wasteland to Vision of Sustainability

Historically, the Western Harbour was a shipbuilding area. With shipbuilding in decline and few other industrial activities present, city leaders recognised by the late 1980s that Malmö's identity as an industrial city lay more in its past than its future. In 1996, the Reepalu administration bought the 175-hectare abandoned industrial area with a vision of creating an eco-district—one that would integrate sustainability principles while relinking Malmö to the sea. The intent was to make Western Harbour a model for the rest of the city and "an internationally leading example of environmental adaptation of a densely built urban environment".[4] An opportunity to realise this goal emerged when Malmö was selected as the site for the 2000 *Swedish Housing Expo*.[5] Inaugurated in 2001, it was called the *European Housing Exhibition: the Bo01 City of Tomorrow* and the neighbourhood called Bo01.

To address mitigation, Bo01 focused on energy efficiency and renewable energy. City planners set an ambitious goal of a total energy consumption of 105 kWh per m^2 per year for construction of 450 apartments—about half of Malmö housing and below Sweden's standard of 110 kWh per m^2 for new construction. Bo01 also runs on 100 % combined renewable energy: solar, wind, geothermal and biogas. To address adaptation, planners created a sustainable urban drainage system (SUDS) that filters rainwater naturally: rain falls on green roofs and down to channels that lead to collection ponds and to canals and swales that lead out to the sea. In this way, urban flooding is limited, particularly important in this coastal community. With these goals in mind, planners invited architects and construction companies to submit plans to develop the parcels. Many were concerned that achieving such a goal would be prohibitively expensive; still 20 building companies and 40 architectural firms accepted the challenge.

One city planner explained that the city planning team, with members from several departments, engaged in ongoing meetings with private sector developers on how to achieve the goals. Standards were established for Bo01 (called the *Quality Programme*) on architectural style, materials, energy, systems technology, green space and SUDS. Architects designed with these standards in mind; construction companies invested in new technology and construction methods to increase energy efficiency.

Nonetheless, Bo01 fell short on achieving energy efficiency standards. Moreover, apartments sell for about twice the national average. Consequently, the planning team reconvened to learn from what happened to improve on the next phase of development.

---

[4]http://www.malmo.se/English/Sustainable-City-Development/Bo01---Western-Harbour.html (accessed 11 August 2011).

[5]Since the 1930s, Sweden sponsors a housing expo circa every 2 years to demonstrate urban best practice.

Applying Lessons and Raising the Bar

Planning for the second phase of Western Harbour, *Flagghusen* (Bo02) started in 2006, incorporating Bo01's lessons learned. One reason few Bo01 buildings met their energy efficiency goals was because it was an expo, with large energy-inefficient windows and the application of new (untested) technologies. City planners wanted to ensure that Bo02's energy efficiency goal was achieved, so standards were raised to 120 kWh per m^2, together with a better electronic system for monitoring building performance so builders would know how systems were performing. Although higher than Sweden's national target, this number is all-inclusive. Likewise, two buildings were built to passive house standards (i.e. designed with highly efficient insulation and materials in order to capture heat energy from, for example, appliances and warm bodies). Remaining energy used for domestic applications cannot exceed 120 kWh per m^2 per year.[6] Affordability was also emphasised, by including units with rent caps. Finally, Bo02 would not benefit from national funding.

During Bo02, Malmö embraced a participatory planning process, the *Building and Living Dialogue*[7] that engaged city representatives, property developers, architects and construction firms in a mandatory series of meetings on energy efficiency, renewable energy, green space planning, safety and affordability. City planners took developers to the best performing buildings in Bo01 and to other developments to examine passive buildings. In addition to testing new approaches in construction and building systems, the Dialogue required competitors to work cooperatively. This wasn't for everyone; of the 39 building companies invited to participate in the development, only 13 did.

The *Building and Living Dialogue* began again for the third phase, *Fullriggaren* (Bo03). Buoyed by Bo02's success, participants agreed to increase standards. A goal was set to have one-third of the buildings meet passive house standards, as established by the newly developed *Environmental Building Programme for Southern Sweden*. This building programme has a three-level building classification system: Grade A uses 55 kWh per m^2 per year (considered passive); Grade B uses 65; and Grade C uses 85. Further, 50 % of the units were to be affordable rentals.

With three phases complete by 2013, Western Harbour has achieved several sustainability goals. On energy, Bo01 is powered by 100 % local renewable energy. On efficiency, most Bo02 buildings meet Grade B standards; Bo03 will exceed this. On green and blue spaces, Bo01 has an SUDS system that includes green roofs, collection ponds and canals. Bo02 and Bo03 do not have SUDS, but emphasise green space (e.g. roof gardens, green walls) to manage rainfall and nurture aesthetic appreciation. On waste management, solid waste is recycled or converted to energy; organic waste produces biogas. On transport, Western Harbour favours pedestrians and bicycles, with good public transportation connections.

---

[6]Other standards apply. See http://www.passiv.de/en/02_informations/02_passive-house-requirements/02_passive-house-requirements.htm (accessed 6 June 2016).

[7]This is part of a national programme (Smedby/Neij, 2013).

The city planning team knew that the ambitious goals and approach to planning applied in Western Harbour also had to work for redeveloping existing neighbourhoods, particularly in lower income areas. The then head of the Environmental Department's Sustainable Development Unit notes, "The city has more existing areas in need of renovation than clean slates". Malmö is redeveloping five city districts to meet sustainability standards, together totalling 312 hectares as part of its *Master Plan*.

The first trial to apply Malmö's sustainability goals in an existing neighbourhood was *Augustenborg*, a predominately immigrant community of 3000 residents with unemployment rates up to 70 %. Built in the early 1950s, the apartments were in need of renovation. This neighbourhood characterises a key challenge faced by the city: Malmö is a magnet for immigrants. Previously, Malmö attracted Greeks, Italians and Yugoslavs to work in shipbuilding and manufacturing. Recently, the immigrant base has expanded, with more than 170 different ethnic groups now living in Malmö. And while this has strengthened Malmö's cultural diversity, it has also brought tensions.

To address challenges, Malmö initiated a participatory planning process, started by *Malmö Municipal Housing Company* and city representatives who met with residents to discuss the redevelopment. City officials focused on energy efficiency and seasonal flooding. Residents also wanted solar panels like the Western Harbour. Although not part of the original plan, they were added in response to resident input. Planners learned how to initiate and implement participatory planning, including how to compromise or clarify decisions with residents, becoming standard practice in other neighbourhoods.

All 1800 apartments were retrofitted, reducing energy consumption by 20 %, and a new passive-energy apartment was built. Storm-water management was addressed via SUDS and by adding 9500 m^2 of green roofs. Innovation was truly local. The particular idea for the storm-water collection system originated in a resident's basement; this later became a new clean-tech company, *Watreco AB*. Residents also created a carpool in 2001; this later became part of the regional non-profit, *Skåne's Carpool*.

An unanticipated outcome was that voting participation increased from 53 to 80 % in early project phases. Trevor Graham comments, "People notice when someone is listening and it makes a big difference in participating in democratic society". Planners and city officials gained residents' trust when they saw that the redevelopment did not lead to gentrification. The social mix remains as before.

The lessons of participatory planning when retrofitting an existing neighbourhood offered valuable insights. Similar strategies were next applied to *Rosengård*, a social housing complex built in the 1960s/1970s during Sweden's *Million House Programme*.[8] Rosengård is an immigrant area (80 %) and has

[8]The Million House Programme (1965–1974) was implemented by the Swedish Social Democratic Party, aiming to build a million new dwellings in a 10-year period, specifically so that everyone afford a home.

suffered from social conflict. Malmö is aiming to redeveloping this troubled neighbourhood, with a focus on environmental technology, increased social and economic integration and better connectivity. Lessons from Western Harbour and Augustenborg are being applied to plan the retrofit.

Malmö is on its way to achieving its climate strategy goals, including for its municipal organisation to become climate neutral by 2020, and the entire city fossil fuel free by 2030. Malmö's success has many elements, *inter alia*: adoption of renewable energy and energy efficiency; a concentration on green and blue spaces; urban planning and transport planning; and participatory planning. All of these components were adopted and adapted in Western Harbour, and adapted again in later developments.

4 Analyzing Malmö's Success

Although not using the term, those we interviewed emphasised how city departments became learning organisations. Malmö is engaged in several national, European and international networks that keep elected officials and planners apprised of effective practices. Our interviewees also discussed how experimentation is key to innovation in planning, and how Malmö's top leadership supports it. Deputy Mayor Anders Rubin comments, "We don't allow anyone not to innovate; we don't say we have done this before. Experimentation is essential to our progress". He credits hiring young innovative thinkers and promoting knowledge sharing among city departments as critical to achieving Malmö's high goals. The planning strategy was to create a clear vision of a sustainable city, test it in an eco-district and apply the knowledge to later developments. Both new planning and technology approaches were tested.

Technology experimentation has become an important part of redevelopment in Malmö. *Sege Park*, a housing/industrial district, is Malmö's test bed for renewable energy and is home to Scandinavia's largest solar photovoltaic array—a 1250-m installation atop a hospital roof, with a maximum production of 166 kWh electricity. This bold installation draws attention and creates dialogue on renewable energy.

Experimentation with new systems has led to new city standards. After testing green infrastructure in Western Harbour, city planners developed a "green space factor" for the entire city. Informed by a similar system in Berlin, the formula allocates points based on how much they contribute to natural drainage and calculates how much land area of a particular property is dedicated to green space [31]. As with energy efficiency, the city sets the initial standard and lets developers figure out how to reach them.

Private-sector actors have also been influenced by experimentation. In Bo02, one building company built two full-size apartment buildings—one with low-energy technology, and the other to a passive-house standard—to compare short- and long-term heating and cooling costs when occupied. The passive house was more cost effective and the company now builds passive houses in other parts of Sweden. Without participating in the *Building and Living Dialogue*, the company would not have attempted this. Several city planners noted that Stockholm officials attempted stricter building standards, but it was more difficult to get building companies on board as they started working with the *Building and Living Dialogue* only later.

The ability of Malmö to become effective learning organisations is directly related to working across administrative silos. City planners commented that collaboration and cooperation among departments is a key factor in achieving so much in so little time. Sustainability is integrated into every department's policy and decision-making; and departments work together to achieve mutual goals.

The Building and Living Dialogue further illustrates how cooperation and collaboration occur with the private sector, creating effective new practices. Trevor Graham notes that this process has planners taking on new roles. "It used to be that planners simply planned, but turned implementation over to developers. Now, city planners are part of the project management team throughout. Because of this relationship, we are able to work together to increase standards".

It can take time to change behaviour, and the *Building and Living Dialogue* played a role in this story. Via continuous dialogue, city planners convinced construction companies to decrease the total number of parking spaces per apartment building, if they emphasised car sharing—more cost effective for construction companies which build costly underground parking spaces for rental to residents. In Bo03, *Sunfleet Car-share* was hired to provide 10–15 cars. Construction companies now struggle to sell the remaining parking spaces, even at the reduced level; residents prefer car-sharing, biking or public transport.

City planners specify two underlying methods for Malmö's achievements. First is an emphasis on inter-departmental dialogue and working across silos, as well as with private-sector actors, to address a particular issue. Second is a continuous process of double-loop learning, with learning from mistakes, acknowledging actual results and revaluating how to reach its goals. In many cases, double-loop learning may highlight effective strategies to apply in other parts of the city. Equally, when goals are not met, strategies are reassessed and lessons are incorporated to move forward accordingly.

Double-loop learning across departments has now become formalised. Project managers from environment and planning departments meet every 4–6 weeks to discuss how to work together and reflect on ongoing projects. Beginning informally, this process became official in 2012 as a way to ensure that obstacles to cooperation are addressed and good practices become institutionalised.

To generate ongoing capacity for sustainability planning, Malmö's local authority and Malmö University also host *The Institute for Sustainable Urban Development* to develop Malmö into a model city for socially and economically sustainable development, by creating a forum for students, researchers and city planners to share knowledge and ideas so new planning processes can be designed, tested and implemented.

Despite efforts and the presence of a large-scale eco-district, this process is far from over. Equally, the decline of the shipbuilding industry had a significant impact on reducing Malmö's emissions—perhaps more than its sustainability planning. More recently, GHG emissions are rising again with the opening of E-ON's natural gas-fired combined heat and power (CHP) plant in 2009, thereby increasing Malmö's GHG emissions for the first time in over a decade, while demonstrating the role industry still plays for Malmö's GHG reduction efforts. It seems that even with the most aggressive sustainability planning, progress is two steps forward, and one step backwards.

5 Conclusion

This chapter examined the role of an eco-district to promote broader climate change planning, using the case of Malmö, Sweden. It asked two questions: (1) Can eco-districts be used by city governments as test beds or learning labs, in particular if or when cities incorporate a deliberate process of organisational learning, allowing city planners and elected officials to apply lessons learned? (2) In response to criticisms that eco-districts are merely islands of green privilege, can these lessons learned be applied in lower income neighbourhoods?

First, we see that elected officials and planners in Malmö have successfully used eco-districts as learning labs, with lessons learned applied to other parts of the city, suggesting that eco-districts have potential for accelerating climate change planning in other cities. Second, in response to criticisms that eco-districts are merely islands of green privilege, we see a deliberate attempt to take the eco-district concept to existing neighbourhoods and to prevent gentrification in so doing. Meanwhile, we recognise that some scholars may view Malmö's efforts as "eco-branding" while highlighting macroeconomic shifts of industry to other global locations, which assisted Malmö's GHG reductions and facilitated its sustainability goals [32]. Nonetheless, we maintain that the city planning process offers valuable lessons for other cities.

Malmö illustrates that meeting climate change goals requires innovative, collaborative and deliberate city planning, including (1) rethinking how the city interacts with the private sector, such as the *Building and Living Dialogue*; (2) exploring how departments plan and work together, such as through formalised regular meetings for this purpose; (3) encouraging experiment to find new ways to meet aggressive goals; and (4) making a concerted effort to include low- to moderate-income housing in all developments and extending the eco-district concept to lower income neighbourhoods. Effective implementation, however, required an ongoing process of double-loop learning, especially as challenges persist before these concepts are fully incorporated—notably in lower income neighbourhoods. And here Malmö can consider itself lucky, having strong and consistent leadership, which encourages such learning. City officials refer to the "long journey" towards sustainable development, indicating that efforts will continue, both in the Western Harbour and in other city developments. Meanwhile, eco-districts allow planners to experiment with new approaches and technologies, enabling shorter implementation cycles and quicker feedback. Planners and elected officials evaluate what works, what does not and why, afterwards applying the lessons to other parts of the city.

Effectiveness is as much about changing the planning process as it is about new technology. Further, the willingness to experiment and to learn from failures inherent to experimentation has allowed Malmö planners to adopt the technologies that work best in the context of the local culture. Without the creation of an eco-district, these learning opportunities would not have occurred.

While few cities may have the level of policy and financial support available to Malmö, yet, any city can replicate the practices of experimentation or breaking down departmental silos to advance sustainability planning practices as demonstrated by Malmo. Malmö's vision of becoming *Sweden's Most Climate-friendly City* was not just a pipedream; it was a step for city planners, across departments, to redefine how they worked and what they could achieve as a result.

References

1. U.N-Habitat: Global Report on Human Settlements 2011 Cities and Climate Change. Earthscan, London (2011)
2. Baeten, G.: Normalising neoliberal planning: the case of Malmö, Sweden (Chapter 2). In: Tasan-Kok, T., Baeten, G. (eds.) Contradictions of Neoliberal Planning, pp. 21–41. Springer Science+Business Media B.V. (The GeoJournal Library, 102, doi:10.1007/978-90-481-8924-3_2) (2012).
3. Fitzgerald, J., Tumber, T., Corley, S.: District-Scale Sustainability Scan. Northeastern University Resilient Cities Lab, Boston, MA (2014)
4. Fraker, H.: The Hidden Potential of Sustainable Neighborhoods. Island Press, Washington, DC (2013)
5. Anders, S.L.: Environmental gentrification in a post-industrial landscape: the case of the Limhamn quarry, Malmö, Sweden. Local Environ. http://dx.org/10.1080/13549839.2013.843510 (2014)
6. Sussman C: Toward the Sustainable City: Vancouver's Southeast False Creek. Master's degree thesis, University of British Columbia. https://circle.ubc.ca/bitstream/handle/2429/43164/ubc_2012_fall_sussmann_cornelia.pdf?sequence=1 (2012). Accessed 17 April 2015
7. Bulkeley, H., Castán Broto, V.: Government by experiment? Global cities and the governing of climate change. Trans. Inst. Br. Geogr. **38**, 361–375 (2012)
8. Lenhart, J., Bouteligier, S., Mol, A., Kern, K.: Cities as learning organisations in climate policy: the case of Malmö. Int. J. Urban Sust. Dev. **6**(1), 89–106 (2014)
9. Argyris, C., Schön, D.: Organizational Learning II: Theory, Method and Practice. Addison Wesley, Reading, MA (1996)
10. Holden, M.: Social learning in planning: Seattle's sustainable development codebooks. Prog. Plann. **69**(1), 1–40 (2008)
11. Schön, D.: The Reflective Practitioner: How Professionals Think in Action. Basic Books, New York (1983)
12. Storbjörk, S.: It takes more to get a ship to change course. Barriers for organisational learning and local climate adaptation in Sweden. J. Environ. Policy Plan. **12**, 235–254 (2010)
13. Beatley, T.: Green Urbanism: Learning from European Cities. Island Press, Washington, DC (2000)

14. Birch, E., Wachter, S. (eds.): Growing Greener Cities: Urban Sustainability in the Twenty-First Century. University of Pennsylvania Press, Philadelphia (2008)
15. Fitzgerald, J.: Emerald Cities: Urban Sustainability and Economic Development. Cambridge University Press, New York (2010)
16. Wheeler, S., Beatley, T. (eds.): The Sustainable Urban Development Reader. Routledge, London (2009)
17. Slavin, M. (ed.): Sustainability in America's Cities: Creating the Green Metropolis. Island Press, Washington, DC (2011)
18. Bulkeley, H.: Urban sustainability: learning from best practice? Environ Plan A. **38**(6), 1029–1044 (2006)
19. Rutherford, J.: Unbundling Stockholm: the networks, planning and social welfare nexus beyond the unitary city. Geoforum. **39**, 1871–1883 (2008)
20. Checker, M.: Wiped out by the 'greenwave': environmental gentrification and the paradoxical politics of urban sustainability. City Soc. **32**(2), 210–229 (2011)
21. Dooling, S.: Ecological gentrification: re-negotiating justice in the city. Crit. Plann. **15**, 40–57 (2008)
22. Shaw, K.: A response to "the eviction of critical perspective from gentrification research". Int. J. Urban Reg. Res. **32**, 192–194 (2008)
23. Argyris, C., Schon, D.: Organisational Learning: A Theory of Action Perspective. Addison Wesley, Reading, MA (1978)
24. Argyris, C., Schon, D.: Theory in Practice: Increasing Professional Effectiveness. Jossey Bass, San Francisco (1974)
25. Dieleman, H.: Organizational learning for resilient cities, through realizing eco-cultural innovations. J. Clean. Prod. **50**, 171–180 (2013)
26. Kolb, D.A.: Experiential Learning: Experience as the Source of Learning and Development. Prentice Hall, Englewood Cliffs, NJ (1984)
27. Flyvbjerg, B.: Five misunderstandings about case-study research. In: Seale, C., et al. (eds.) Qualitative Research Practice pp, pp. 420–434. Sage, Thousand Oaks, CA (2004)
28. Yin, R.: Case Study Research: Design and Methods. Sage, Thousand Oaks, CA (2003)
29. Germanwatch and Climate Action Network Europe: The Climate Change Performance Index 2011. Germanwatch and Climate Action Network Europe, Brussels (2011)
30. Granberg, M., Elander, I.: Local governance and climate change: reflections on the Swedish experience. Local Environ. **12**(5), 537–548 (2007)
31. Kruuse, A.: The Green Space Factor and the Green Points System. Town and Country Planning Association and European Union Green and Blue Space Adaptation for Urban Areas and Eco Towns Network, London (2011)
32. Anderberg, S., Clark, E.: Green sustainable Oresund region: ecobranding Copenhagen and Malmö. In: Vojnovic, I. (ed.) Sustainability: A Global Urban Context, pp. 591–610. Michigan State University Press, East Lansing, MI (2013)

Index

S. Dhakal, M. Ruth (eds.), *Creating Low Carbon Cities*,
DOI 10.1007/978-3-319-49730-3

MIX
Papier aus verantwortungsvollen Quellen
Paper from responsible sources
FSC® C105338

If you have any concerns about our products,
you can contact us on
ProductSafety@springernature.com

In case Publisher is established outside the EU,
the EU authorized representative is:
Springer Nature Customer Service Center GmbH
Europaplatz 3, 69115 Heidelberg, Germany

Printed by Libri Plureos GmbH
in Hamburg, Germany